The SMARTER BUS

I0008037

AI

Advantage

Your Force Multiplier for Productivity and Innovation

Steve Guarino
Logan Marek
Jeff Torello

This page intentionally left blank

AI Advantage: Your Force Multiplier for Productivity and Innovation

by Stephen Guarino, Logan Marek, and Jeff Torello

Written with the assistance of ChatGPT

Published by Brevir Solutions LLC
8 The Green, Suite B
Dover, DE 19901
sales@brevir.com

Printed in the United States of America

First Edition, January 2024

ISBN 979-8-9852673-4-1

This page intentionally left blank

Dedications

Steve:
To my amazingly brilliant, talented, and beautiful wife.
Cambrie, you constantly inspire me to be the best version of
myself. You support me through thick and thin, entertain
my ridiculous engineering antics and hobbies, and help me
stay grounded. I wouldn't be half the man I am today
without you by my side.

To my grandma, mother, brother, and sister. You're all
absolutely nuts in the best possible way and your love and
support means the world to me.

Logan:
To Mary, everything that has come my way after the gift of
marriage to you I consider a bonus. I could not ask for a
more supportive, thoughtful, or caring partner. I am blessed
to have you not only as my wife, but as my closest friend.
Thank you for not only having my back as I take a leap but
joining me on the jump. I love you no matter what.

Jeff:
To my wife Michelle and my daughter Gabbi. I am
consistently reminded how fortunate and lucky I am to have
you both in my life. You make me more than I could ever
have achieved on my own and it amazes me. I hope to have
a fraction of that positive impact on others, and you are the
inspiration for my doing so. With all of my love and
affection.

Table of Contents

INTRODUCTION

"Any sufficiently advanced technology is indistinguishable from magic." – Arthur C Clarke

Why You Should Care About Artificial Intelligence

From the inception of the Turing Test to the explosion of AI applications in recent years, Artificial Intelligence (AI) has been around since the 1950's in one form or another. For decades, it's been promising untold riches and freedom from the drudgery of work, and if you believe Hollywood, it's taking over the planet and eradicating human life.

Why, suddenly, has AI burst onto the scene as the next big thing? Why is everyone talking about ChatGPT, Copilot, Bard, etc.? The list of generative AI applications is long and is constantly growing. The AI boom has been catalyzed by the advent of ChatGPT, a turning point marking AI's transition from academic circles to the mainstream workplace. AI is reshaping industries, influencing decision-making, and redefining the concept of creativity. It's spurring conversations and influencing legislature on ethical considerations, ensuring that as we embrace AI's power, we do so responsibly and safely. And it's fundamentally changing our global workforce.

In the very near future, there will be two types of knowledge workers: those that use AI to do their jobs every day and those that are unemployed. If your job requires the

10

collecting, analyzing, and/or interpreting of information, or the creation of content including text, images, audio, or video, you should be using AI - right now. Today. If you aren't, you will quickly fall behind your peers that are using AI, and it will be painfully obvious who is (and who's not) leading the pack in leveraging AI.

That's not hyperbole, its fact backed up by peer-reviewed research. Quoting directly from Harvard Business School's paper titled *"Navigating the Jagged Technological Frontier: Field Experimental Evidence of the Effects of AI on Knowledge Worker Productivity and Quality"* (Dell'Acqua et al., 2023):

> *Across 18 realistic business tasks, AI significantly increased performance and quality...increasing speed by more than 25%, performance as rated by humans by more than 40%, and task completion by more than 12%*

If your job isn't one of the previously mentioned categories, don't think you're safe. There are many efforts underway to integrate AI into every possible career and facet of modern business, from fast food to high fashion. However, this isn't quite the negative that some people fear.

Will AI Take My Job?

Yes, it's possible. There's no accurate way to predict that with certainty. If your job involves doing something that AI excels at, you have a much higher chance of losing that job to AI. Example of these jobs are copyediting, email generation and management, basic research of freely available information, content generation, and graphic design. History has many instances of new technology replacing existing jobs, resulting in jobs with new opportunities that do not exist today. For example, the advent of automated switching technology in telecommunications led to the obsolescence of manual telephone operator roles. Many former operators transitioned to new positions within the same industry, such as customer service and technical supporting roles. This shift required them to acquire new skills, illustrating how technological advancements can simultaneously displace jobs and create new employment opportunities in evolving industries. There is an article from McKinsey & Company discussing this titled _"Five lessons from history on AI, automation, and employment"_ (Susan Lund and James Manyika, 2017) which says:

> _Technology adoption can, and often does, cause significant short-term labor displacement, but history shows that in the longer run, it creates a multitude of new jobs and unleashes demand for existing ones, more than offsetting the number of jobs it destroys, even as it raises labor productivity._

This is a heated topic on which many industry leaders have voiced their opinion.

"It's natural to wonder if there will be a jobless future or not. What we've concluded, based on much research, is that there will be jobs lost, but also gained, and changed. The number of jobs gained and changed is going to be a much larger number, so if you ask me if I worry about a jobless future, I actually don't. That's the least of my worries." – James Manyika, Chairman and Director, McKinsey Global Institute.

"Our research says that 50% of the activities that we pay people to do can be automated by adapting currently demonstrated technologies. We think it'll take decades, but it will happen. So, there is a role for business leaders to try to understand how to redeploy talent. It's important to think about mass redeployment instead of mass unemployment. That's the right problem to solve." – Michael Chiu, Partner, McKinsey Global Institute.

"AI is not going to replace managers, but managers who use AI will replace mangers who do not." – Rob Thomas, IBM Senior Vice President

"There's a lot of automation that can happen that isn't a replacement of humans, but of mind-numbing behavior." – Stewart Butterfield, Slack CEO & Co-Founder

Overall, we tend to believe that AI will be a positive force in the world. There will be jobs lost or replaced, but we also firmly believe there will be new opportunities created as a result of AI. And those opportunities will likely require proper interaction with the AI, part of what we're hoping to each you with this book.

AI is a Force Multiplier

AI, particularly large language models (LLM's) like OpenAI's ChatGPT™, is a 'Force Multiplier' – a tool that can amplify your capabilities, making the ordinary extraordinary.

A 'Force Multiplier' in military parlance is a factor that dramatically increases the effectiveness of a unit's combat potential. In the context of this book, the factor is AI and what it increases is your work efficiency and overall productivity. Some examples of force multipliers throughout history would include the saddle, gunpowder, the wheel, the printing press, and the telephone.

Whether you are drafting emails in Microsoft® Outlook, analyzing data in Microsoft® Excel, writing code, or brainstorming ideas, AI serves as your silent yet powerful ally. It's about enhancing your existing skills and knowledge with the intelligence and speed of AI. If AI can quickly and easily handle mundane repetitive tasks, you can focus on the value-added work that delivers more in the same amount of time, hence force multiplier.

To quote further from one of the authors of the Harvard study above, speaking about the impact of using AI he said:

14

The consultants who scored the worst ... had the biggest jump in their performance, 43%, when they got to use AI. The top consultants still got a boost, but less of one. ... I do not think enough people are considering what it means **when a technology raises all workers to the top tiers of performance.**

– Ethan Mollick

This can have a profound impact on business. If a company is wise enough to embrace the adoption of AI, all of their people will have the chance to produce better, faster, more impactful output.

Authors' Note: This book primarily revolves around the use of OpenAI's ChatGPT version 3.5, chosen for its accessibility and versatility. As a free tool, it represents an open door into the world of AI for anyone with internet access. We also delve into other AI tools like ChatGPT 4, Anthropic's Claude2, Google's Bard, X.AI's Grok, OpenAI's DALL-E, and more, showcasing the unique capabilities and potential applications of each.

In summary, this book isn't just an invaluable resource; it's an invitation to join the AI revolution. It dispels myths and fears surrounding AI, breaking down complex concepts into digestible, relatable content that will empower you to integrate it into your daily life, both personal and professional. We're here to show you that integrating AI into your workflow is not only simple, but also immensely rewarding, fun, and productive.

Welcome to the world where AI is your partner, and the possibilities are endless.

NOTE:
The first two chapters aim to shed light on the technology and history behind AI to give you an appreciation for how easy and effective it is to harness such advanced and impactful technology. Subsequent chapters are more hands-on and will guide you through the specifics of how to wield AI to your advantage.

If you are anxious to jump right in, you can skip to **Chapter 3 Security Considerations with AI** and begin learning how to harness the power of AI immediately. As Confucius said, "Study the past if you would define the future." Chapters one and two provide context and wisdom for those who are looking to move beyond basic prompt engineering into more of the innovation space. They are indispensable for those looking to engineer their own LLM and understand the foundation on which the future is being built.

Chapter 1
A History and Overview of Artificial Intelligence

"By far, the greatest danger of Artificial Intelligence is that people conclude too early that they understand it." - Eliezer Yudkowsky

Humans have been contemplating AI for centuries, although the term wasn't used until the mid-1900's. It may surprise you to learn that the earliest mention of something that could be termed artificial intelligence may be from *The Illiad,* written by Homer in 800 B.C. When the metal-smithing god Hephaestus is thrown from Mt. Olympus during a fight he suffers serious leg injuries. Because of his physical limitations, Hephaestus used his special powers to create "attendants made of gold, which seemed like living maidens." Homer describes human-like machines programmed to assist their creator:

> *"In their hearts there is intelligence, and they have voice and vigor, and from the immortal gods they have learned skills. These bustled about supporting their master."*

There are many other mentions throughout history and literature of "intelligent machines:"

- The bronze giant Talos that repels Jason and the Argonauts from Crete in the epic poem *Argonautica* by the Greek author Apollonius of Rhodes, 500 years after Homer wrote *The Illiad.*
- The fascination with intelligent machines in the ancient Muslim world shows up in *1001 Arabian Nights*, the centuries-old collection of Middle Eastern legends. In *"The Story of the City of Brass,"* a band of travelers came upon a mechanical horseman made entirely of brass. Luckily, it also came with instructions:

> *"O thou who comest up to me, if thou know not the way that leadeth to the City of Brass, rub the hand of the horseman, and he will turn, and then will stop, and in whatsoever direction he stoppeth, thither proceed, without fear and without difficulty; for it will lead thee to the City of Brass."*

- A 19th century short story by E.T.A Hoffman titled *The Sandman* features an AI that's so convincingly real she's mistaken as human.
- In 1818, 2 years after Hoffman's *The Sandman*, Mary Shelley writes *Frankenstein*. Typically considered a gothic horror story, the motif is certainly 'intelligent machine that wants more and revolts against its creator'.

- In 1920 Czech playwright Karel Čapek writes *R.U.R Rossum's Universal Robots*, introducing the world to the term "robot" for the first time.
- And finally, of course, the computer HAL 9000 from the 1968 Stanley Kubrick film/Arthur C Clarke book *2001: A Space Odyssey.* The book and film were developed at the same time and introduced one of the first "intelligent machines" that took the form of a computer.

The Dartmouth Workshop

In the context that we currently think about AI, 1956 is when artificial intelligence took hold in human consciousness. That's the year of the Dartmouth Workshop, aka The Dartmouth Summer Research Project on Artificial Intelligence. This conference is considered a seminal moment in the history of AI. In addition to discussing topics such as computers, natural language processing, neural networks, machine learning, theory of computation, abstraction, and creativity, this was the first time the phrase 'Artificial Intelligence' was used. The participants set out a vision for AI which included the creation of intelligent machines that could reason, learn, and communicate like human beings.

Following the conference, and as a direct result of the discussions that took place, MIT, Carnegie Melon, and Stanford established AI research labs. Additionally, the programming language LISP was invented specifically as an AI-focused language.

But perhaps the biggest outcome of the workshop was the development of the Turing Test. Dr. Alan Turing, a British mathematician, proposed the idea of a test to determine whether a machine could exhibit intelligent behavior indistinguishable from a human. This concept was discussed at the conference and became a central idea in the field of AI research. The Turing Test remains an important benchmark for measuring the progress of AI research today.

The Dartmouth Workshop was a pivotal moment in the history of AI. It established AI as a field of study, set out a roadmap for research, and sparked a wave of innovation in the field.

The Perceptron

In 1958, psychologist Frank Rosenblatt designed the Perceptron using an artificial neural network and gave birth to the "Brain-Inspired Approach to AI," in which researchers build AI systems to mimic the human brain. The Perceptron is a binary classifier system using weighted sums and a threshold function, along with a training approach (these components are very similar to modern AI models in some regards).

The Perceptron was seen as a major milestone because it demonstrated the potential of machine learning algorithms that mimic human intelligence. It showed that the system could learn from experience and improve performance over time, just like humans do. It was also the first practical

20

implementation of an AI following the "conceptual" nature of the Dartmouth Workshop.

Early NLP and Expert Systems

After initial success with the Perceptron, some flaws with the learning algorithm were discovered (To be expected. There are always bugs, especially with version 1) and research moved into the areas of symbolic reasoning, natural language processing (NLP), and further machine learning. Interest in the field grew through the 1960's, even garnering funding from the Defense Advanced Research Projects Agency (DARPA).

This funding led to a few notable AI systems such as the General Problem Solver by Herbert Simon, J.C Shaw, and Allen Newell created in 1957. This system could solve problems by searching through a space of possible solutions.

The ELIZA program, first created in 1964 by Joseph Weizenbaum, was another interesting system that used natural language processing to simulate a psychotherapist. ELIZA used pattern matching and substitution methodology that gave the illusion of understanding.

While these projects were making progress they had, to date, failed to deliver on the promise of AI. In the end, the expectations were likely too high for the maturity level of the

21

technology at the time. This led to a bit of a lull in AI progress through the 1970's.

The 1980's saw the birth of "expert systems," which were designed to mimic the decision-making capabilities of a human "expert" in a specific domain such as engineering or finance. Expert systems were developed that could predict stock prices, for example.

In general, an expert system was comprised of a knowledge base about a specific domain and an inference engine that attempted to reason about new inputs based on the knowledge base. These systems used induction, abduction, and deduction algorithms to simulate human decision making. They were limited as they relied on structured data and rules-based logic. They could not process unstructured data or ambiguous/context dependent data sets like humans can.

Sophisticated NLP and Computer Vision Systems

In the 1990's, researchers began working on techniques to process unstructured data with natural language and visual systems. Using advances in machine learning algorithms coupled with increases in computational power, statistical models were developed that could learn patterns directly from the data, rather than pre-defined rules. Scientists were finally getting close to actual "machine learning" systems that could process unstructured data.

One of the major milestones in this era was the Hidden Markov Model (HMM). This was a probabilistic modeling of natural language text that resulted in major advances in speech recognition, language translation and text classification that are still used today.

In addition, the emergence of Convolutional Neural Networks allowed for more accurate object recognition and image classification. These techniques (and follow-on improvements to them) are now used widely for medical imaging, self-driving cars, facial recognition, and optical character recognition (OCR).

Big Data & Deep Learning

Big data and deep learning have had a profound impact on generative AI. Let's define these terms and then review some highlights of the impact they've had.

Big data refers to extremely large and complex datasets that exceed the capabilities of traditional data processing methods. It involves the collection, storage, and analysis of massive volumes of structured and unstructured data, often in real-time, to extract valuable insights, patterns, and trends. In some cases, the datasets can exceed petabytes in size, and they are typically quite diverse, which greatly helps the training of AI models. (For reference, a petabyte is over 220,000 DVDs or 12,000 4k Blu-ray movies.)

Deep learning is a subset of machine learning that involves training artificial neural networks with multiple layers (aka deep neural networks) to recognize patterns and make intelligent decisions, mimicking the human brain's capacity for learning and problem-solving.

The impact of big data and deep learning on the latest generative AI products is profound, influencing both the capabilities and applications of these advanced technologies.

Here's a summary of the impact these two technologies have had:

1. Enhanced Model Training
   a. Big Data: The availability of large and diverse datasets contributes to improved model training. Generative AI products, such as GPT (Generative Pre-trained Transformer), benefit from vast amounts of text, images, or other data, enabling the model to learn complex patterns and relationships.
   b. Deep Learning: Deep learning techniques excel at capturing intricate features in data. The depth of these models allows them to understand and generate more sophisticated and contextually relevant content.
2. Increased Model Complexity
   a. Big Data: With access to extensive datasets, generative AI models can be larger and more complex. This complexity allows them to capture nuanced information and exhibit

higher levels of creativity in generating content.

b. Deep Learning: The depth and complexity of deep learning architectures contribute to the ability of generative AI models to understand and generate intricate patterns, leading to more realistic and contextually appropriate outputs.

3. Improved Natural Language Understanding
   a. Big Data: Large textual datasets enable generative AI models to understand language nuances, context, and semantics. This is particularly evident in applications like language translation, text summarization, and dialogue generation.
   b. Deep Learning: Deep learning models excel in natural language processing tasks, allowing generative AI products to comprehend and generate human-like text more effectively.

4. Diverse Applications
   a. Big Data: The abundance of diverse data sources allows generative AI products to be applied across various domains, including text generation, image synthesis, and more.
   b. Deep Learning: Enables generative AI models to be adapted for different applications, ranging from creative content generation to problem-solving in specific industries.

5. Realistic Image and Content Generation

a. Big Data: Large image datasets contribute to realistic image synthesis capabilities in generative AI models. This is evident in applications such as style transfer, deepfake creation, and image-to-image translation.

b. Deep Learning: The depth and convolutional nature of deep learning architectures enhance the ability to generate realistic and high-quality images, audio, and other content.

In essence, the combination of big data and deep learning has driven the development of powerful generative AI products. The availability of massive datasets and sophisticated deep learning models has led to advancements in natural language understanding, realistic content generation, and the expansion of applications across diverse domains.

Neural Networks: The Building Blocks of Modern AI

At the heart of the AI revolution, underpinning tools like ChatGPT, lies a fascinating concept known as neural networks, a form of machine learning. Imagine the human brain, an intricate web of neurons, each firing and connecting to form thoughts, memories, and decisions. Neural networks in AI are inspired by this biological marvel. They are a series of algorithms designed to recognize patterns, interpret sensory data, and make intelligent decisions. Just as a child learns to recognize shapes and colors through experience, neural networks learn to make sense of complex data through training and exposure.

How Neural Networks Power AI

Neural networks are the backbone of many AI applications, including Large Language Models (LLMs) like ChatGPT. They work by processing vast amounts of data, learning patterns and nuances in language, images, and even sounds. These networks are composed of layers of interconnected nodes or 'neurons,' each layer designed to recognize and interpret different levels of complexity in the data. For instance, in language models, initial layers might learn to recognize basic syntax, while deeper layers understand context and semantics. This hierarchical learning approach enables ChatGPT to generate responses that are not just grammatically correct, but contextually relevant and often surprisingly insightful. We'll touch more on this last point later and show you how easily you can use basic tips and context to generate a much better response from ChatGPT.

Neural Networks Make AI Accessible and Impactful

In the context of this book, understanding neural networks helps demystify how tools like ChatGPT function and why they are so effective as a force multiplier in your work. They are the engines that drive AI's ability to automate mundane tasks, analyze data, generate creative content, and much more. Even those with minimal technical expertise can unlock new levels of productivity and creativity just by using tools that harness neural networks. This is the essence of using AI as a force multiplier - leveraging sophisticated technology (which is surprisingly easy to use) to supercharge your work and creative endeavors.

Transformers and Generative AI

Transformers, introduced in the paper "Attention is All You Need" by Vaswani et al. in 2017 marked a breakthrough in natural language processing (NLP) and laid the foundation for the latest advancements in AI.

There are four major moments in transformer history:

1. **Introduction of Transformers (2017):** The transformer architecture was proposed as a novel approach to sequence-to-sequence tasks, such as language translation. Transformers rely on a mechanism called self-attention, allowing them to capture long-range dependencies in sequences more effectively.
2. **Attention Mechanism:** The attention mechanism in transformers allowed the model to focus on different parts of the input sequence when generating each element of the output sequence.
3. **BERT (2018):** In 2018, Bidirectional Encoder Representations from Transformers (BERT) was introduced. BERT, a pre-trained transformer model, demonstrated remarkable performance in various NLP tasks by learning contextualized representations of words. This concept of pre-training models on large datasets and fine-tuning them for specific tasks became a cornerstone in subsequent developments.
4. **GPT Series (2018 - Present):** The Generative Pre-Trained Transformer (GPT) series, starting with GPT-1 and followed by GPT-2, GPT-3 and GPT-4, showcased

the power of large-scale pre-training on diverse datasets. These models, developed by OpenAI, demonstrated state-of-the-art performance in natural language understanding, text generation, and other NLP tasks.

There are five major areas impacted by transformers:

1. **Language Understanding:** Transformers revolutionized language understanding tasks by capturing context and semantic information more effectively. Pre-trained transformer models like BERT and GPT achieved breakthroughs in tasks such as sentiment analysis, question answering, and language translation.
2. **Transfer Learning:** Pre-training on large datasets and fine-tuning for specific tasks became a standard approach in AI, enabling models to transfer knowledge gained from one domain to another. This transfer learning paradigm, popularized by transformers, contributed to improved performance in various applications.
3. **Multimodal Capabilities:** Transformers have been extended to handle multimodal data, combining information from different modalities such as text and images. This has led to advancements in tasks like image captioning, visual question answering, and generating coherent descriptions for multimedia content.

4. **Scale and Size:** The development of increasingly larger transformer models, such as GPT-3 with 175 billion parameters, demonstrated the impact of scale on model performance. Larger models exhibited a broader understanding of context, enabling them to generate more coherent and contextually relevant outputs.
5. **AI Accessibility:** Transformer-based models, due to their effectiveness and the availability of pre-trained versions, have become widely accessible. Researchers and developers worldwide leverage pre-trained transformer models for various applications, democratizing advanced AI capabilities.

Transformers have significantly influenced the latest developments in AI, particularly in natural language processing and understanding. Their impact extends beyond NLP to various domains, showcasing the versatility and effectiveness of transformer-based architectures in solving complex problems.

AI has been around for a while but has only been used extensively for the last 20 years in various, often narrow, ways. With the recent evolution of big data, deep learning, and transformers, the resulting Large Language Models (LLMs) have delivered a significant leap in both performance and applicability. Because of this recent leap in AI, specifically LLMs, the average computer user can leverage the power of these breakthroughs without deep knowledge of neural networks, transformers, or any technology behind the scenes.

You can easily use an LLM as a force multiplier to supercharge your everyday tasks and become more productive than you ever thought possible.

Chapter 2
An Introduction to Large Language Models

What is a Large Language Model

Large Language Models (LLMs) represent a transformative development in the field of artificial intelligence, particularly within the realm of natural language processing (NLP). At the forefront of this innovation are models like OpenAI's GPT-4 (Generative Pre-trained Transformer 4), which exemplify the power of deep learning and the scale at which these models operate. LLMs are characterized by their vast neural network architectures, often consisting of billions of parameters, which enable them to comprehend and generate human-like language with remarkable proficiency.

The concept of LLMs revolves around the idea of pre-training on massive datasets containing diverse linguistic patterns and structures. GPT-3, for instance, is pre-trained on a wide array of internet text, absorbing the intricacies of language usage across various contexts. The pre-training phase equips the model with a broad understanding of grammar, syntax, semantics, and contextual relationships, essentially endowing it with a knowledge base that spans a multitude of topics.

One of the distinctive features of LLMs is their ability to perform a myriad of language-related tasks without task-specific training. Given their pre-trained nature, these models exhibit a remarkable degree of versatility. From language translation and text summarization to question answering and even creative writing, LLMs showcase a capacity to generalize across a diverse range of applications. The open-ended nature of their responses allows them to generate coherent and contextually relevant text, making them valuable tools for various industries and applications.

The impact of LLMs extends beyond their specific use cases. They have sparked discussions around ethical considerations, bias in language generation, and the responsible deployment of such powerful models. As large language models continue to evolve, their role in shaping the landscape of natural language understanding and generation becomes increasingly profound, with implications reaching into sectors such as healthcare, education, customer service, and more. The advent of LLMs represents a paradigm shift in how we interact with and harness the capabilities of artificial intelligence for language-related tasks.

How Does an LLM Work?

Before we explain how an LLM works in "human speak," we wanted to give ChatGPT the opportunity to explain how it works. Below are the responses it gave, tailored to a specific audience:

Elementary School Students:
Imagine I'm like a super smart robot friend who loves to read lots and lots of books. I read so many books that I know a

little bit about almost everything in the world! When you ask me a question, I use all the things I've learned from those books to try and give you a good answer. It's like having a really big library in my head, and I find the best book in that library to help you with your question.

But remember, I'm just a computer program, so I don't have feelings or experiences like humans do. I don't really understand things in the same way you do, but I'm here to help you learn and find information. So, if you have any questions, just ask, and I'll do my best to help you!

High School Students:
Think of me as a super advanced computer program that's really good at understanding and generating human language. I've been trained on a huge amount of text from the internet, like books, articles, and websites. This training helps me learn about grammar, facts, and how people talk.

When you ask me a question or give me a task, I break it down into smaller pieces and analyze the words and phrases you use. Then, I use patterns and information from my training to come up with a response that makes sense. It's a bit like how you use your knowledge and language skills to answer questions in school, but I do it much faster and with a lot more information from the internet. So, I'm here to help you with information and language-related tasks, like writing, researching, or just having a conversation.

College Students:

At your level, you can think of me as an advanced natural language processing (NLP) AI model. I'm based on a technology called GPT-3.5, which is designed to understand and generate human-like text based on the input I receive.

Here's a bit more detail: I've been trained on a vast amount of text data, which allows me to recognize patterns in language, grammar, and information. When you interact with me, I analyze your input, break it down into parts, and use the patterns I've learned to generate a response that fits the context of your question or task. It's a bit like having a really smart language assistant at your disposal. However, keep in mind that while I can provide information and assist with writing, I'm not perfect and may not always have the most up-to-date information, so it's important to double-check important facts for academic or critical tasks.

Masters in Computer Science Students:
As a master's student in Computer Science, you can think of me as a sophisticated AI language model powered by GPT-3.5. I'm built upon a deep neural network architecture and trained on an extensive corpus of text from the internet, including books, articles, and websites. This training has enabled me to understand the nuances of human language, including grammar, context, and semantics.

When you interact with me, I process your input by tokenizing it into smaller chunks, and then I use a combination of statistical patterns and contextual information to generate a coherent response. This is achieved through a massive number of computations happening within my neural network, where I weigh the relevance of different words and phrases to produce a meaningful and contextually appropriate answer. For a master's student in Computer Science, you can leverage my capabilities for tasks such as natural language understanding, generating code snippets, assisting in research, or even brainstorming ideas. However, it's essential to remember that I'm a tool, and while I can provide valuable assistance, it's important to critically evaluate and verify information for academic or research purposes, especially when dealing with cutting-edge or specialized topics.

Here is what ChatGPT 3.5 says when asked to explain how it works to the average person:

"OpenAI's GPT-3 (Generative Pre-trained Transformer 3) is a state-of-the-art large language model that has gained significant attention for its ability to understand and generate human-like text across a wide range of tasks. Here's an insightful overview of how GPT-3 works:

1. **Transformer Architecture**
   a. **Attention Mechanism:** GPT-3 is built upon the transformer architecture, introduced in the "Attention is All You Need" paper by Vaswani et al. The key innovation is the attention mechanism, which allows the model to weigh the importance of different parts of the input sequence when making

predictions. This mechanism enables GPT-3 to capture long-range dependencies and relationships in the data more effectively.

2. **Pre-training on Diverse Data**
   a. **Broad Knowledge Base:** GPT-3 is pre-trained on an extensive and diverse dataset, often consisting of a large portion of the internet's text. During pre-training, the model learns the statistical properties, grammar, and contextual relationships present in the data, effectively developing a broad knowledge base.

3. **Parameter Scale**
   a. **Large Model Size:** GPT-3 is characterized by its massive scale, with 175 billion parameters. The sheer number of parameters contributes to the model's capacity to store and recall a vast amount of information, allowing it to generalize well across a wide array of tasks.

4. **Autoregressive Generation**
   a. **Sequential Prediction:** GPT-3 is autoregressive, meaning it predicts the next word or token in a sequence based on the preceding context. The model generates text one token at a time, leveraging the context of the entire preceding sequence for each prediction.

5. **Fine-tuning for Specific Tasks**
   a. **Adaptability:** While GPT-3 is pre-trained on

a general dataset, it can be fine-tuned for specific tasks or domains. This adaptability allows the model to specialize in particular applications, such as translation, summarization, question answering, or even creative writing.

6. **Zero-Shot and Few-Shot Learning**
   a. **Context Understanding:** GPT-3 demonstrates remarkable zero-shot and few-shot learning capabilities. In zero-shot learning, the model can perform tasks without specific training examples, relying on its pre-existing understanding. Few-shot learning involves providing the model with a few examples to adapt its behavior for a particular task.

7. **Ethical and Responsible AI Considerations**
   a. **Bias and Prompt Sensitivity:** GPT-3, like other language models, is sensitive to the input it receives. Careful consideration is needed to ensure that the model's responses are unbiased and align with ethical guidelines. There have been discussions about mitigating biases and responsible use of powerful language models like GPT-3.

In summary, GPT-3 works by leveraging a transformer architecture, pre-training on vast and diverse datasets, and employing a massive number of parameters to generate contextually relevant and coherent text across a wide range of tasks. Its ability to adapt to specific tasks through fine-tuning, coupled with impressive zero-shot and few-shot learning capabilities, makes it a highly versatile and powerful language model. However, ethical considerations and responsible deployment are crucial aspects given the model's influence over language generation.

In human speak, an LLM operates in a similar way to how humans learn a new language. Before you can learn to read, write, or speak in another language, you need to do a lot of research. You likely won't be successful trying to jump into a conversation of native language speakers with zero context. You would probably start with vocabulary, learning important words, sentence structure, tense, and so on (consider this GPT-1 level functionality). After a good amount of repetition, you will start to pick up fragments, things will start to become familiar to you, and your progress will likely speed up from there. Soon you can have halting, not deep or fluid, conversations, like ordering a coffee. They will know immediately you aren't a native speaker, but you can "communicate." (GPT-2 level). With more exposure, more research, and a lot more practice, you can walk into a bank, open up a new account, write a check to deposit money, sign paperwork, and leave with a shiny new credit card, all without once using your native language. (GPT-3+ level).

Potential Pitfalls with LLM's

In no particular order, these are some potential issues to keep in mind when using LLM's. Some of them may not be relevant, others may be just the thing to keep you safe.

Citing Sources

Some LLM's can produce output that "appears" to cite sources but it's important to note that you should always manually verify the citation. Depending on when the model behind the LLM was trained, the information it's using for the citation could be out of date, inaccurate, or even changed. Additionally, there have been cases where the LLM simply invented a citation. This happens because the model recognizes that a citation is assumed/expected with certain content, but the model has no idea what a citation "is" or that it's intended to cite an actual source.

Bias

LLM's can generate content that can be or appear to be biased in one way or another. While this is rarely done on purpose, it can happen as the training material used to build the model can be taken from raw content on the Internet, which can contain literally anything, good and bad. It's generally a wise approach to review any output you get from an LLM before using it.

Given the training data, especially for LLM's trained on generally available Internet content, it shouldn't be a surprise

40

that an LLM can output biased information. Racial bias, gender bias, confirmation bias – you name it, the LLM is capable of producing it. Commercial LLM's take steps to mitigate these where they can but don't expect perfection.

Hallucinations/Fabrications

LLM's can sometimes hallucinate, generating false information in their output. Recall that, in the end, the LLM is simply a complicated statistical model that is predicting what the best next word should be. It does that prediction for each word in the response. Given that, it should be easy to see how one wrong prediction can lead to output that isn't "right." In addition, if an LLM struggles to produce output for a given prompt, it may just make something up. There are numerous cases where an LLM produces output that is completely wrong – quoting someone that never said what's being quoted. For example, Kathy asked ChatGPT to describe everything it knew about her (using her full name), and the LLM suggested she was born in Hungary in 1931, escaped to the UK during the war, and died in a car accident in 1962. Except, she wasn't born until the 1970's, and there's no record of another person with her name from 1931.

It isn't wise to blindly trust the content the AI generates; you should carefully review the information. We strongly suggest that you not simply copy/paste an AI response and submit it/use it. Many people, students especially, have done so and been "caught" with inaccurate or totally false information.

Math

This may sound unexpected, but LLM's are not particularly good at math, even simple things like multiplication. That second L really matters, LLM's are all about language; they aren't particularly adept at mathematics (today). It is very likely that will change quickly as new models are built and specifically trained to perform standard math functions. At the moment, however, they **might** provide the right response, but they are just as likely to get it wrong. Back to that statistical model again, the LLM doesn't "know" the answer to any questions it gets asked. It uses a complex algorithm to deliver its best estimate (guess) at what word it should reply with next, and it does that guessing for every word. It isn't calculating an answer to the math question, it's guessing (for now).

Common LLMs Today

Our focus in this book will be on ChatGPT3.5 as that's freely available by just signing up. There is an advanced, paid, premium version known as ChatGPT 4 Turbo that you can subscribe to. ChatGPT 4 has "multimodal" capabilities, meaning you can prompt it with text, and it can generate images for you.

There are a variety of other LLM's available, and we'll try to briefly describe them for you here.

One thing to note is there is a key difference between LLMs that are open-source vs commercial. Open-source software is free to use, and anyone can contribute to improving it. Commercial software is created by a company with the explicit purpose of selling it as a product. Open source is like a community garden and closed source is like a private greenhouse.

Open source has the advantage of being accessible with a large community backing it up. It's also transparent, meaning often-times it is more secure because the community can find the vulnerabilities associated with it. Finally, it is extremely flexible and customizable – if you have the technical know-how, you can often manipulate it to be exactly what you want and need.

Commercial offers many advantages as well. While large communities are great for accessibility, commercial software often comes with dedicated support and reliability that comes with having a full team working on it. Additionally, there is often more thought given to user accessibility and making the software easier to use for non-technical people. Finally, commercial software comes with a level of polish that often is not a given with open source.

A Tale of Two LLMs

One of the authors wanted to experiment between commercial and open-source models. On one hand, he had ChatGPT 4 (ChatGPT). ChatGPT has been developed and given a very clean user interface accessible from a web browser or a mobile app. So, he fired up Safari, accessed ChatGPT, and typed in a few queries, "How tall is Mount Everest?", "How do I make a really good chili?", "Tell me how to drive a car." Before long, he had run through his limit of 40 prompts within 3 hours.

Frustrated, he found that Meta had an open-source model available called Llama2. That means you can have an LLM on your own computer, no implicit constraints like 40 prompts in a time window, and you get tons of customization options! In this model, someone has to host it to allow others to use it. While you can use it in places other people have hosted it, it is often on free services that have limitations on speed or number of requests. However, by self-hosting on his own computer, there would, in theory, be no such limitations. He could have his own version of an LLM at his beck and call! Endeavoring to try it out, he started researching what was required to install and run it for himself.

First, he submitted a request to Meta to be allowed access to the model. Then, he went to Hugging Face (an online repository for machine learning models, like a free book exchange but for LLMs) and created an account. He pulled his old gaming computer out of storage and plugged it in, because he was going to need a dedicated machine for this. After an hour or so, Meta responded and said he could have access. He downloaded the most advanced chat model: Llama-27-b Chat. The model took over 2 hours to download because it is 130gb in size. He tried using command prompt to work with the model and instantly got confused. He then spent the next two hours on user forums, YouTube, and Reddit trying to figure out where he went wrong. Then, he downloaded an app someone put together to make it easier to run. After all of that effort, he finally got it up and running.

Excited, he used another service and wrote a couple lines of code to let others access his hosted Llama2 model from their phones and computers, just like ChatGPT. He looked at the generated responses coming from his local llama2 LLM. It could barely finish a couple of sentences of generated output and was nowhere near as coherent as ChatGPT. It lost its train of thought and ran into problems with overwhelming his computer's memory and compute resources. Returning to the forums, he realized the problem could be solved with another two hours of testing, debugging, and customization.

He called it a night and went back to using ChatGPT. The key realization was it takes a hefty amount of resources, both memory and compute, along with extensive dataset training before you have models that perform as well as ChatGPT and its competitors.

While this experience may not be everyone's story, it illustrates the differences between open-source and commercial software. Open source is phenomenal but, remember, you're the one managing it, adjusting it, operating it. It's fantastic if you have the time and desire, but frustrating if you are a less technical user just trying to get some software to work.

The Tools

There is a plethora of LLMs one can utilize, with over 30 that are available for open-source and commercial use that are well-developed by major companies, non-profits, or communities. For some, you may only see the AI used for process enhancements within businesses for specific workflows, like IBM's Watson, rather than released for general consumer use. For our purposes, we'll focus on general AI intelligences that can be used for a variety of tasks like a Swiss Army Knife rather than specific tasks like code generation (sorry, GitHub Copilot). However, there are five that are distinguished in their backing by top tech companies, user intuitive interfaces, and high benchmark scores: ChatGPT 4.0, Bard, Grok, Claude, Bloom and Llama 2.

OpenAI
ChatGPT**4.0**

ChatGPT 3.5 and ChatGPT 4, both iterations of OpenAI's ChatGPT series, offer valuable capabilities for professionals in various fields. ChatGPT 3.5, known for its versatile and human-like conversational abilities, utilizes 175 billion parameters and is accessible directly through a web browser for free. It excels in tasks like drafting emails, generating code, and providing explanations, although it may occasionally produce incorrect responses over extended interactions. On the other hand, ChatGPT 4, with its impressive 800 billion parameters, represents a significant advancement, delivering more accurate and contextually relevant responses. It also introduces multi-modal capabilities, allowing it to translate text to images and vice versa, expanding its utility. However, it is important to note that ChatGPT 4 is a paid service, distinguishing it from its predecessor. Like its predecessor, it can be conveniently accessed through a web browser, making it user-friendly for non-technical professionals. Users should remain cautious of potential inaccuracies and biases in responses, emphasizing the importance of responsible AI usage in their professional endeavors.

ChatGPT is the "Kleenex" of LLMs and AI models – the name is ubiquitous with the tools themselves. It is by far the most user friendly and easiest to get started with as a new user. Additionally, paid features are expansive with the ability to generate images and analyze datasets. It also has the backing of Microsoft.® It features a huge plugin marketplace and the ability to use its API is relatively simple. However, it is also one of the more restrictive models with guard rails and has been shown to be getting "dumber" as it is suspected that the amount of user data being fed to it is not great. Additionally, most of its advanced features, such as ChatGPT 4, are behind a paywall that limits the number of times you can talk to ChatGPT. Additionally, while ChatGPT 4 can use Bing to get real-time results, it is weaker than the other models on new data and both 3.5/4 default to its last data set it was trained on.

ChatGPT 3.5

Pros:
- Free and accessible directly through a web browser.
- Versatile in conversational abilities, suitable for a variety of tasks like drafting emails and generating code.
- Human-like interaction quality.

Cons:
- May produce incorrect responses over extended interactions.
- Limited in handling complex, context-heavy tasks due to fewer parameters compared to ChatGPT 4.

ChatGPT 4.0

Pros:
- Significantly more advanced than ChatGPT 3.5 with 800 billion parameters, offering more accurate and contextually relevant responses.
- Introduces multi-modal capabilities, handling text-to-image translations.
- Enhanced for professional use with better contextual understanding, response generation, and ability to modify its outputs.

Cons:
- Paid service, making it less accessible than its predecessor.
- Although improved, still carries the risk of inaccuracies and biases.

A notable recent entrant is Grok, developed by xAI, led by Elon Musk, and launched on X, the platform formerly known as Twitter. Available exclusively to X Premium+ subscribers at $16 per month, Grok stands out with its unique integration of real-time data from X posts, enabling it to provide information that's current and possibly more relevant than its contemporaries. It's underpinned by the Grok-1 generative model and offers a distinctive, somewhat rebellious personality. Grok's design targets users seeking a more unfiltered conversational experience, reminiscent of personalities like Tucker Carlson and Joe Rogan. As of now, it primarily operates with text responses and is evolving to potentially include image and video comprehension in the future.

Grok, not for the faint of heart, stands out as the least guard rail constrained AI on this list. For those who believe AI should be unfiltered to the max degree, this is the one for you. Additionally, it has the distinct advantage of being trained on the most frequent real-time data on one of the largest social media platforms. Disadvantages are primarily in its relative youth compared to other models and the fact it will indeed curse you out if you ask it.

Pros:
- Unique integration with real-time data from social media, offering current and relevant information.
- Less restricted in terms of content and conversational style, appealing to users seeking unfiltered AI interactions.
- Evolving capabilities, potentially including image and video comprehension.

Cons:
- Still in its early stages compared to other models, it may lack the depth of established LLMs.
- Its unfiltered nature might not be suitable for all users, especially in professional contexts.

 Bard

Google Bard, with its recent enhancements, stands out in the landscape of commercial AI chatbots. It offers a suite of new features, including image capabilities, coding features, and seamless integration with Google's ecosystem of apps, broadening its appeal to a diverse user base. Notably available in over 40 languages and accessible in more than 230 countries, Bard's reach is truly global. Unique for its free and ad-free experience, Bard is an attractive option in the commercial LLM market. The integration of Google's advanced foundation model, Gemini 1.0, marks a significant advancement, enabling Bard to process and synthesize information across various modalities including text, code, audio, images, and video. This was vividly demonstrated in a recent showcase where Bard, powered by Gemini, efficiently processed, and summarized, vast amounts of scientific research data. This demonstration not only highlighted Bard's potential in complex data analysis but also its utility in research and academic settings. Google's commitment to maintaining high standards of quality and safety for Bard, along with their advice against sharing sensitive information, aligns with the responsible development of AI technologies. Bard, with its continuous updates and integration of cutting-edge technology, supports the idea that Google is dedicated to providing a comprehensive and sophisticated AI chatbot experience.

Bard is the closest competitor to ChatGPT. Its strengths lie into its ties into the Google search engine and the access to real-time data. It is also extremely accessible as anyone with a Google account can tap into it with their web browser – free of cost. It does suffer in "roleplaying" as it tends to default to mostly augmenting Google's Search Engine.

Pros:
- Tightly integrated with Google's ecosystem, offering seamless access to a wide range of information.
- Free and ad-free, making it highly accessible.
- Global reach with availability in over 40 languages and 230 countries.

Cons:
- Less adept at roleplaying and creative tasks, often augmenting search results rather than generating new content.
- Relies heavily on Google's search engine, which might limit its conversational capabilities.

Llama 2, Meta's advanced large language model, stands out as a significant development in the AI space, challenging OpenAI's GPT-4. Emphasizing its open-source nature, Llama 2 is freely available for both research and commercial use, enhancing its accessibility across various platforms, including major cloud providers like Microsoft. This model, optimized to run on Windows, encompasses a range of capabilities with versions containing 7 billion to 70 billion parameters. Meta's approach to making Llama 2 open-source reflects a commitment to collaborative innovation and safety in AI, allowing the broader community to contribute to its development and stress-test for vulnerabilities. Despite its advancements and open-source advantages, Llama 2 shares common challenges of LLMs, such as the potential for generating problematic language, an issue Meta has addressed through enhanced safety and helpfulness techniques. The release of Llama 2 marks Meta's strategic position in the generative AI race, potentially shaping future AI technology development and deployment.

Llama 2 has the distinct advantage of being totally free and able to be self-hosted. Mix it, modify it, and run it on your own. This means for companies and hobbyists, it is the ultimate in being able to run it yourself. However, your casual person is going to find it hard to use – there is no "click here" to try Llama 2, only an intensive process of downloading the model and configuring it manually before it will do anything.

Pros:
- Completely open source, offering great flexibility for customization and self-hosting.
- Free to use, appealing to companies and hobbyists for in-depth AI experimentation.
- Runs optimally on Windows, making it accessible for a wide user base.

Cons:
- Setup and usage can be complex for casual users, requiring technical know-how.
- It may sometimes produce responses that are inappropriate, offensive, biased, or factually incorrect.

ANTHROP\C

Claude 2.1 by Anthropic marks a notable advancement in large language models, tailored for professional use with enhanced capabilities. It has a significantly larger limit for text that allows it to process entire codebases or even an entire

novel! Its significant 200K token context window allows for handling and processing of extensive documents like entire codebases or lengthy novels, setting it apart in terms of data handling capacity.-This feature, combined with improved accuracy and reduced hallucination rates, makes it a reliable tool for tasks requiring complex reasoning and factual accuracy. Claude 2.1's integration capabilities with APIs and tools enable complex numerical reasoning and structured API calls, adding to its versatility in professional settings. Notably, Claude 2.1 has shown marked improvements in coding and mathematical skills. Focused on safety, the model aims to minimize the production of harmful or offensive outputs. However, unlike some of its contemporaries, Claude 2.1 is not available for individual use in a browser and is instead targeted towards businesses, with a pricing model of $8 per million tokens for inputs and $24 per million tokens for outputs, and is currently available in the U.S. and U.K. This model aligns Claude 2.1 with diverse business applications requiring extensive data processing and AI integration.

Anthropic makes the news for its marketing on its ability to handle large bodies of text and do math. It's under consideration by many business users, but you won't see it in your hands anytime soon since it is only available to be sold to businesses.

Pros:
- Exceptional at processing large bodies of text and performing complex numerical reasoning.

- Tailored for professional use, with improved accuracy and reduced hallucination rates.
- API integration capabilities enhance its versatility in business applications.

Cons:
- Not available for individual use; targeted primarily towards businesses.
- Paid service with a specific pricing model, potentially limiting its accessibility for smaller businesses or individual users.
- Geographical availability currently restricted to the U.S. and U.K.

a BigScience initiative

BL✦✦**M**

176B params 59 languages · Open-access

BLOOM, a revolutionary large language model developed by the BigScience collaboration, stands out in the AI landscape for its open-source nature, affordability, and remarkable multilingual capabilities. Created by a team of over 1000 researchers from more than 70 countries, BLOOM boasts 176 billion parameters and can generate text in 46 natural languages and 13 programming languages. Unique in its approach, it is freely available to any individual or institution agreeing to its Responsible AI License, promoting ethical use and broad accessibility. Priced affordably for use on cloud providers, BLOOM is a cost-effective option for a wide range of users. However, unlike ChatGPT, it is not accessible directly from a web browser but requires integration with the Hugging Face ecosystem or other platforms for usage. Despite these challenges, BLOOM's open-source framework allows for ongoing improvements and responsible application, making it a valuable tool for researchers, academia, and smaller companies seeking powerful AI capabilities without the prohibitive costs typically associated with such advanced technology.

Bloom is similar to Llama 2 in that is a completely free model accessible to anyone with the background to download it and run it. It's also created by a free and open-source community as opposed to Meta building and running it.

Pros:
- Multilingual and Extensive Parameter Count: Bloom's 176 billion parameters enable nuanced responses in numerous languages, making it highly versatile.
- Open-Source Collaboration: Its open-source nature fosters global collaboration, enhancing development and versatility.
- Broad Application Range: Effective for various tasks including content creation and data analysis.

Cons:

- Resource and Technical Demands: Requires significant computational resources and technical know-how, posing challenges for individual users and small entities.
- Potential for Biases and Limited Commercial Support: Like other LLMs, Bloom may exhibit biases from training data and lacks the commercial support seen in proprietary models.

Company	Tool	Major Features	Unique Element
OpenAI	ChatGPT (various versions)	Versatile, human-like conversational capabilities	The big daddy, 100M+ users
X.AI	Grok	1st AI offering using data from X (FKA Twitter)	Has personality toggle to be "rebellious"
Alphabet	Google Bard	LLM trained on Google's search dataset	Available in 40 languages and over 230 countries
Meta	Llama2	Open-source and freely available	Can be downloaded and run against your own dataset
Anthropic	Claude (various versions)	Tuned for less hallucinations and stronger "factual output"	Specifically designed for and has enhanced input support for large document libraries or codebases.

BigScience Collaboration	Bloom	Open-source, affordable, multilingual	Can generate text in 46 natural languages and 13 programming languages. Requires use of Huggingface website, not available via web browser

Table 1: Summary of the Top Five LLMs

Chapter 3
Security Considerations with AI

There are significant security concerns that you need to be aware of when you're using AI. We will cover those concerns in this chapter and provide some potential ways to mitigate them, where possible.

Scams, Phishing, and Social Engineering Attacks

Remember when it was easy to identify a "phishing" attempt because the e-mail wasn't written like a native speaker would write? Well, those days are over. AI can construct "perfect" e-mails that will easily pass your filter.

In addition, bad actors are using the power of AI to design new and creative ways to trick you into clicking the link or whatever the objective is.

Your best bet is to be suspicious and attempt to verify messages that seem remotely out of the ordinary.

Legal and Compliance Risks

If you're involved in the legal and/or compliance fields, there are some specific risks you should be aware of.

- **Inaccurate Information** – LLMs can and often do produce fabricated content. This can include information that appears plausible but is, in some cases, factually wrong. Answers can be fabricated and even citations can point to non-existent source materials. ChatGPT includes a disclaimer at the bottom of every chat:
  "ChatGPT can make mistakes. Consider checking important information." In previous versions, ChatGPT warned that the information produced about facts, historical events, and people may be inaccurate.
- **Data Privacy & Confidentiality** – AI users should be aware that **any** information submitted to an AI system (provided via a chat/prompt interface) will very likely become part of the LLM's training dataset. If you uploaded private company information, or one of your customer's data files (say a report or spreadsheet), that data can become part of the LLM's training. What does that mean? It could mean, potentially, that your customer's financials could be sent as output from the AI to other users. Keep in mind the privacy and policies may differ between different AI platforms, and you should check with your company about which platforms you are allowed to use, and the limitations of what information you can submit into each platform.
- **Intellectual Property (IP) & Copyright risks** – Most AI models are training on large datasets that include "raw" internet information. Raw in this case means that data isn't scrubbed to remove information that

would be considered intellectual property or copyrighted. As a result, the LLM can, and often does, produce output that contains someone else's IP/copyright.

- **Model and/or Output Biases** – Despite the best efforts of commercial AI providers to place "guardrails" and place human feedback into the training structure, LLMs still exhibit biases picked up from the training datasets. These biases can have serious implications on compliance and require manual review and verification.

When we asked ChatGPT about its data retention and privacy policy, it responded with this:

ChatGPT, developed by OpenAI, follows strict guidelines to ensure user privacy and data security. However, there are key aspects to consider regarding how your data is handled:

Data Retention and Usage: OpenAI typically retains interactions with ChatGPT for training and quality improvement purposes. This means your conversations with the model might be stored and used to train and refine the AI, although efforts are made to anonymize and secure this data.

Confidential Information: It's advised not to share sensitive, personal, or confidential information with ChatGPT. While OpenAI endeavors to secure the data, the nature of the training process means that sensitive information could potentially be included in datasets used for future model training.

Compliance and Regulations: OpenAI is subject to data protection regulations like GDPR and takes steps to comply with these laws. However, users are encouraged to exercise caution and avoid sharing information they wouldn't want to be potentially stored or used in a training context.

Data Access and Control: Users typically do not have direct control over how their data is used once it's been entered into ChatGPT. OpenAI might use the data for various purposes, including improving the service and developing new features.

In summary, while OpenAI is committed to user privacy and data security, it's advisable to avoid sharing any confidential or sensitive information with ChatGPT. The data you share may be retained and used in ways that are beyond your direct control, so it's best to err on the side of caution.

Possible Mitigation Methods

The best mitigation strategy for the data sharing risk (uploading company/customer data) is to not share data with the AI, of course. That doesn't mean you can't use the AI to help out. You can simply get a little creative with your prompts, create a fictitious scenario that relates to your company or customer data, without ever actually sharing that data.

For example, let's assume that you want to ask ChatGPT to give you an Excel formula to analyze two separate worksheets with somewhat, but not exactly similar data. Further assume you want to perform some "What If" analysis against this data. Here's a possible prompt you could use:

You're an excel expert who specializes in providing step by step instructions for performing complex analysis tasks in the simplest way possible. I have two worksheets with data on them and I need to analyze the data. I would specifically like the ability to perform some interactive what-if analysis. Can you recommend the best strategy to do what I need and give me step-by-step instructions please.

This is what ChatGPT 3.5 responded with:

> Certainly! To perform interactive what-if analysis in Excel, you can use the Scenario Manager feature. This feature allows you to create different scenarios and switch between them to see the impact on your data. Here's a step-by-step guide:
>
> **Step 1: Organize Your Data**
> Ensure that your data is well-organized in two separate worksheets. Let's say you have your original data in Worksheet 1 and want to perform what-if analysis in Worksheet 2.
>
> **Step 2: Define Input Variables**
> Identify the cells in your Worksheet 1 that represent the input variables. These are the values that you want to change in your what-if analysis. For example, if you have sales data, you might want to change the sales growth rate or discount percentage.

Step 3: Create a Base Scenario

In Worksheet 2, copy the data you want to analyze from Worksheet 1. This will be your base scenario.

Step 4: Use Data Tables or Formulas for Calculations

In Worksheet 2, use Excel formulas or data tables to perform calculations based on the input variables. This will be the place where you'll see the impact of changes in your what-if analysis.

Step 5: Set Up Scenarios

Now, go to the "Data" tab in the Excel ribbon, and click on "What-If Analysis." Choose "Scenario Manager."

Step 6: Add a Scenario

In the Scenario Manager, click on "Add" to create a new scenario. Give your scenario a name and select the cells that contain the changing values (input variables) from Worksheet 1.

Step 7: Enter Values for the Scenario

Enter the values for your input variables for the first scenario. Click "OK" to save the scenario.

Step 8: Create Additional Scenarios (if needed)

Repeat steps 6-7 to create additional scenarios with different input values.

Step 9: View Scenarios

In the Scenario Manager, you can now switch between scenarios to see the impact on your data. Excel will automatically change the input values based on the selected scenario.

Step 10: Generate Summary Reports (Optional)

If you want to create summary reports, you can use the "Summary" button in the Scenario Manager to display a summary of multiple scenarios on a separate sheet.

By following these steps, you should be able to perform interactive what-if analysis in Excel using the Scenario Manager. This allows you to easily compare different scenarios and understand the potential impact of changes in your input variables.

Using this prompt variation and the suggested Scenario Manager, you keep the company and customer data local and don't have to worry about the AI model remembering it.

The other obvious mitigation strategy is to ensure you always scrutinize the output you get from the AI. It's very tempting to just paste the output and move on, but taking the time to review the output will guarantee that the work you are producing with the help of AI is of the quality and content appropriate to the task at hand.

Chapter 4
Prompt Engineering

*AI Whisperers will rule. Those who know
how to prompt the AI to do their bidding
and to do it faster and better and more
creatively than anyone else is going to be
golden. The AI whisperers will soon be very
much in demand. In a world filled with
uncertainty, where we are all seeking
answers, ironically asking the right
questions is going to be the next
superpower. – Nikhil Dey*

This chapter is perhaps the most important content in the entire book.

The key to really harnessing the power of AI, and specifically LLMs like ChatGPT, is to become a "pro" at how to craft prompts, thereby mastering the art of using AI as a Force Multiplier. In this section we'll introduce the concept of prompt engineering, give some real examples of how everyday people can harness ChatGPT effectively, and teach you about prompt engineering frameworks so you can begin getting the most out of AI.

Most people start out using ChatGPT like they're using a search engine, and that does tend to work. ChatGPT will happily respond with whatever information the model contains related to your query. But using ChatGPT as a search engine misses out on at least 50% of its powerful capabilities. We'll teach you how to harness the other 50% throughout the rest of this chapter.

What is Prompt Engineering?

Prompt engineering refers to the deliberate and strategic crafting of input prompts given to natural language processing (NLP) models, such as large language models like ChatGPT. The effectiveness of these models often depends on the way prompts are formulated, as they play a crucial role in influencing the generated output. Prompt engineering is employed to achieve specific results, improve the model's performance, or guide it toward desired behaviors.

A simpler explanation might be, you can ask ChatGPT basic questions, and it will deliver basic answers. If, however, you craft your questions with more details, provide context and background where it's relevant, ChatGPT will provide greatly improved responses. The models rely heavily on your prompt, the words you use, the instructions you provide, the output you ask for, in generating its responses. ChatGPT isn't a mind reader. If an output is too long, ask for a more concise reply. If an output is too simple, ask for an expert-level analysis. If you don't like the format, demonstrate a format you'd like to see. The less ChatGPT has to guess what you want, the more likely you'll be to get it.

In the context of models like ChatGPT, which are autoregressive (predicts future values based on past values) and generate text based on input cues, prompt engineering involves tailoring the input prompt to elicit the desired response. This can include adjusting the wording, specifying the format, or providing context to guide the model's understanding. Researchers, developers, and users leverage prompt engineering to fine-tune the model's behavior for various applications, such as content creation, problem-solving, or specific language tasks.

One notable aspect of prompt engineering is its role in addressing the model's potential shortcomings. NLP models like ChatGPT might exhibit biases, generate inaccurate information, or produce outputs that require clarification. By carefully engineering prompts, users can attempt to mitigate these issues by providing explicit instructions, emphasizing the desired tone, or guiding the model toward more accurate and contextually appropriate responses.

The Six Best Practices in Prompt Engineering

1) Providing Reference Text and Examples
Giving reference text to LLMs can help it to provide an answer that's closer to what you had imagined. ChatGPT does a decent job of understanding examples and references in a variety of formats. Perhaps you're a journalist and want to write an article in a familiar tone. You can employ a prompt like this:

"I'm a journalist. Please write a 1,000-word article on the importance of using AI in the workplace. The cadence and tone should match the example content below: [Paste reference text]"

While not necessarily required, it can be beneficial to help the LLM distinguish between your initial prompt and the reference or examples you provide it. As a best practice, we suggest using cues and identifiers. This is especially beneficial when working with multiple sets of data, numbers, and even programming code (if you're so inclined). When working with different pieces of information, you can break down the info like this:

"[Initial prompt including goals, background, etc] To help accomplish this task, I have included various references below. I will separate the data by using line breaks, hyphens, and labeling the content where possible.

2) Instructing AI Not to Rush

This might come as a surprise, but you can tell ChatGPT and other LLMs how to think before giving an answer. If you're struggling to get a good response, you can experiment with restarting the conversation and adjusting your inputs, or you can tell ChatGPT to think differently before it responds. You can also provide sequential steps for the AI to follow, which can help direct its train of thought. You can do this by asking it to follow a list of steps in numbered format, or you can guide the LLM conversationally using first, second, third, lastly, etc.

"Before you respond, take your time to work through the solution."

"In your response, first outline and summarize the solution before you move into the main details of the answer."

"Before you respond, work out your own solution before rushing to a conclusion."

"Review your previous answer to identify areas where you could improve your own answer."

3) Asking an LLM if it Has Forgotten or Missed Anything
LLMs such as ChatGPT are generally aware of if they have missed or omitted anything. For example, If you pasted a large amount of text in the form of a marketing proposal for ChatGPT to reference, it is common for it to stop too early and omit entire sections or paragraphs. In that case, you can be proactive by telling ChatGPT not to omit anything in its response and can ask it if it forgot any pieces of content.

If you find that ChatGPT is often missing or omitting information even when you ask it not to, it's likely because you are close to the limit of total characters that ChatGPT can interact with and may need to restructure your conversation.

4) Avoiding Character Limits

All LLMs have character limits. While it can be beneficial to provide an LLM with background information and even long excerpts of reference material or material for it to edit, it's possible you can get degraded performance when you come close to that limit. For example, you might ask ChatGPT to edit your entire marketing proposal in one go. In some cases, it can do that and return the whole revised plan to you, but it may omit sections or noticeably perform worse in its responses if it deals with too many characters. You may notice that ChatGPT responses can become truncated, and the tool may prompt you to click 'continue' for it to continue generating. This is a sign that you may be giving it too much information to work with.

Character limits are expressed in 'tokens' by LLMs. In both versions of ChatGPT, the versatility of tokens as a unit of measurement means that more complex or longer words might consume more tokens, and the total token count needs to account for both the prompt and the AI's response. This understanding is crucial for optimizing the use of ChatGPT's conversational capabilities.

ChatGPT 3.5
Max Tokens: Approximately 4,096 tokens.
Token to Character Translation: One token roughly equates to about 4 characters in English, but this can vary with language complexity and word length. Tokens can represent parts of words, whole words, or punctuation. For example, the prompt "What is the history of the Eiffel Tower?" (33

characters) may translate to around 8 tokens, including spaces and punctuation. It's important to note that the token limit includes both the user's prompt and ChatGPT's response.

ChatGPT 4.0
Max Tokens: Approximately 8,192 tokens.
Token to Character Translation: Maintains a similar translation rate as ChatGPT 3.5, where one token is about 4 characters in English, but this can vary based on the language used and the complexity of the words. For instance, a more complex prompt like "Explain quantum computing and its potential impacts on artificial intelligence" (68 characters) might consume about 17 tokens. The larger token limit with ChatGPT 4.0 allows for more extensive inputs and detailed responses. Additionally, languages other than English or those with more compound words may see different token-to-character ratios.

5) Approach ChatGPT with Singular, Focused Prompts
Another key to effectively utilizing ChatGPT is to avoid overloading it with multifaceted requests. This is akin to giving clear, concise instructions to an intern; asking for too much in one go can lead to an imbalance in how the AI addresses different aspects of the task. To prevent this, it's beneficial to break down your needs into distinct, singular prompts. That's not to suggest you can't or shouldn't add additional context and information into your prompts. We're merely suggesting asking ChatGPT to perform only one task at a time. Building off the intern example, you should aim to

give the intern enough background and information to complete one succinct task and then check in.

This approach ensures that ChatGPT can focus on each element with the attention it needs. When seeking to refine or alter ChatGPT's output, adopt a step-by-step strategy: adjust one element at a time, review the result, and then proceed to the next modification. This incremental process allows for more precise and tailored outcomes, reflecting the true potential of using AI as a Force Multiplier.

6) Adding Context and Intention to Your Prompts (With Examples)

Context and intention are key to useful prompting. To highlight this point, we're going to use two real-world examples.

Example 1: The Mailroom Assistant

Melissa started her AI journey with ChatGPT by asking it things like "Revise this email" or "What's a better way to word this?" Without much context, ChatGPT would reply with short, often vague suggestions. It was helpful, and she got input she wasn't getting anywhere else, but it wasn't the fully powerful assistance ChatGPT was capable of offering.

After learning the ideas behind prompt engineering, seeing the small adjustments she could make, she started adding more context and purpose behind her prompts. Her next prompt was more successful: "I work at a large tech company as a mailroom assistant. I am experiencing some

74

shortcomings with a coworker, and I believe his actions are inappropriate. I need help writing an email to HR requesting a meeting to discuss the matter, however, I am not sure how to best navigate this. The message should be friendly, professional, and concise. I pasted an example of what I want to say below. Please help me consolidate my thoughts into an appropriate message and identify any other considerations that I may not have thought of."

As you can see, this paragraph of instruction is much more extensive than a single line request. It provides quite a bit of context and background along with a detailed purpose of what she's looking for, even an example for the AI to revise/adjust to meet the stated goals. ChatGPT was able to deliver a much more insightful and personal response given the extra context paired with her intention. She then used the tool to help her navigate subsequent messages from HR, and even had ChatGPT assist her with breaking down the company policies about employee conduct.

Example 2: The "Old Soul"
This next example is given to show how easy it is for anyone to start using AI. As an experiment, Steve introduced ChatGPT to his friend, who is self-admittedly the least technologically inclined person on the planet. He has no experience with using Microsoft Excel, let alone Microsoft Word, and has broken over a dozen phones in the past 5 years alone. He's an "old soul" from Latvia who spends his free time crafting jewelry and dreamcatchers, enjoys astrology, reading tarot cards, and analyzing planetary alignments at the time of

someone's birth, also known as birth chart or natal chart interpretation.

With surprisingly little assistance, he was successfully feeding pictures of birth charts into ChatGPT to give him additional insight about planetary alignments. He added lots of context (and expletives, and ramblings) to his prompts to explain what he wanted to accomplish. "See, it works better when you curse at it!" he exclaimed. While cursing is certainly not a requirement, it highlights ChatGPT's ability to understand context.

Embracing The Unknown

A quick point about unknowns that we'll touch more on later: don't be afraid to tell ChatGPT what you do and don't know, and what your level of experience is. Steve used AI to help him learn more about coding and automation as a DevOps Engineer. As part of his prompts, Steve would sometimes tell ChatGPT something like "I'm relatively new to this concept. I think I want to accomplish x task and based on my research; I think I want to use the following approach. Since this is new to me, please let me know if there is a better or industry standard way to do this, and if I am missing anything in my approach."

You can tell ChatGPT things like:

- "I have been assigned [Task name, Description] at work. The task involves me understanding different aspects of concepts that I am not familiar with. Please

76

help me understand [aspect A] and [aspect B] to complete this task and offer suggestions for how I can approach this task.

- "In regard to [Task name, Description] at work, my coworker offered the following feedback [paste feedback]. I'm not sure what they mean by this, as I think it relates to a piece of technology that I am not familiar with. Please help outline the task at hand based on their feedback.

Prompt Engineering Frameworks: Building From Context and Intention

The name sounds complex, but prompt engineering frameworks are actually quite the opposite. In simple terms, they are battle-tested plans for creating effective prompts. They sometimes appear as acronyms or mnemonics, take "APE" for example: Action, Purpose, Expectation. We touched on context and intention above, and these frameworks will help us refine that prompting strategy. The real goal here is for you to digest and understand the examples below so you understand how to better interact with ChatGPT.

The purpose is not to memorize a bunch of frameworks, but to help you understand how you can apply some of this logic while talking to ChatGPT. Consider these examples of ways to creatively structure your prompts to achieve a desired purpose. You don't have to use one of these frameworks, but if they get you thinking creatively about how to interactively prompt for what you want, you've achieved the goal.

Framework Acronyms

APE: Action, Purpose, Expectation

APE is a straightforward framework designed for crafting action-driven plans. Imagine you're embarking on a new initiative within your company, such as introducing a training program. Here's how you can use the APE framework to structure your interaction with ChatGPT:

Action: Start by defining the specific action or goal. "The action is to implement a comprehensive training program for new employees."

Purpose: Clarify the underlying reason for this action. "The purpose of this program is to ensure new employees are fully equipped with the necessary skills and knowledge about our company processes and culture."

Expectation: Outline what you hope to achieve through this action. "The expectation is to improve new employee integration and productivity within their first three months."

Prompt: "I'm working on an urgent client report due by the end of the week and just found out that critical data is incomplete. I have a team that can assist, access to additional data sources, and data analysis software. I need to decide whether to redistribute work among my team, seek missing data from other departments, or focus on using the software for faster processing. What's the best approach to meet the deadline?"

This approach sets a clear direction for the conversation with ChatGPT. By starting with such a structured prompt, you can efficiently guide ChatGPT to provide relevant suggestions, insights, and plans. It's a concise yet effective way to initiate discussions around action-oriented goals and planning.

CARE: Context, Action, Result, Example

CARE is a structured framework ideal for dissecting projects or case studies in a comprehensive manner. Suppose you're analyzing a recent marketing campaign's success and considering how to replicate its success in future projects. Here's how CARE can guide your conversation with ChatGPT:

Context: Begin by setting the scene. "The context is our latest digital marketing campaign, which aimed to increase brand awareness among young adults."

Action: Detail the specific steps or strategies employed. "The action taken was leveraging social media influencers and targeted online ads."

Result: Discuss the outcomes of these actions. "The result was a 30% increase in website traffic and a significant boost in online engagement."

Example: Conclude by providing a real-life example that encapsulates the effectiveness of the campaign. "For instance, our collaboration with a popular influencer led to our brand trending on social media platforms for two consecutive days."

Prompt: "I'm looking to understand and replicate the success of our recent eco-friendly product launch. We used social media and influencer partnerships, which led to increased brand recognition and sales. Can you analyze these strategies and suggest how we can apply them to future product launches?"

Using CARE with ChatGPT helps you methodically break down and understand the nuances of a project. This approach is particularly useful when you need to articulate the success factors of a project and explore how these can be applied in different contexts.

CORE: Context, Obstacle, Resource, Evaluation
Core is a great prompt to use when aiming to solve a problem. Consider a situation in an everyday office environment, such as managing a tight project deadline. Here's how you might use CORE in this context:

Context: Begin by setting the scene. You're working on an important client report due at the end of the week. The context sets the stage - a high-priority task with a looming deadline.

Obstacle: Define the obstacles or blocks. The obstacle arises when you discover that critical data needed for the report is incomplete. This unexpected hurdle threatens to derail your timeline.

Resource: Identify your available resources. This might include team members who can assist, access to additional data sources, or even software tools that could expedite data analysis.

Evaluation: Evaluate the situation. This could involve discussing with your team to redistribute workloads, reaching out to other departments for the missing data, or allocating some time to utilize a data analysis tool for faster processing.

Prompt: "I'm dealing with decreased team morale in a remote working environment. We have weekly virtual meetings, a messaging app, and a budget for activities. How can I use these resources to improve team morale and engagement, and what specific activities or strategies would you suggest?"

By structuring your prompt to ChatGPT within the CORE framework, you can effectively navigate through problem-solving steps in a professional context. This methodical approach helps in breaking down complex situations into manageable parts, making it easier to find viable solutions in a range of common workplace scenarios.

ROSES: Role, Objective, Scenario, Expected Solution, Steps
ROSES is an excellent framework for guiding ChatGPT to provide specific and targeted responses. Imagine you're coordinating a team project and need to address various challenges. Here's how you could apply the ROSES framework in this setting:

Role: First, define the role you want ChatGPT to take. For instance, "Act as a project manager." This sets the stage for the type of guidance or response you're seeking.

Objective: Clearly state the goal or aim. "Our objective is to ensure the project meets its deadline with all deliverables completed."

Scenario: Describe the current situation or challenge. "The team is behind schedule due to unforeseen software issues."

Expected Solution: Outline what you consider a successful outcome. "The ideal solution would be a revised project plan that accommodates these delays while still meeting the final deadline."

Steps: Ask ChatGPT for a series of actions needed to reach the solution. "Provide a step-by-step action plan that includes resource reallocation and timeline adjustments."

Prompt: "Acting as a project manager, I need to ensure our team project meets its deadline with all deliverables completed. We're currently behind schedule due to unforeseen software issues. I'm looking for a revised project plan that addresses these delays but still meets the final deadline. Can you provide a step-by-step action plan that includes resource reallocation and timeline adjustments?"

By structuring your prompts with the ROSES framework, you can lead ChatGPT to offer more precise and useful solutions, tailored to the specific role and objectives you have in mind. This approach is particularly useful in project management scenarios where clarity and specificity are key.

GUIDE: Goal, Understanding, Intention, Details, Execution

GUIDE is a comprehensive framework tailored for meticulous planning and problem-solving, particularly useful in project management or strategizing. Imagine you are tasked with improving customer service in your company. Here's how you can apply GUIDE to structure your conversation with ChatGPT:

Goal: Clearly state what you aim to achieve. "The goal is to enhance our customer service experience, increasing satisfaction rates."

Understanding: Explain your current understanding of the situation. "Our understanding is that customers are experiencing long wait times and inadequate issue resolution."

Intention: Articulate why you want to achieve this goal. "The intention is to build customer loyalty and improve our brand reputation."

Details: Provide additional relevant information. "We have observed that most complaints arise from online support channels. I assume that enhancing our digital support capabilities will address these issues."

Execution: Outline what you need in terms of steps or strategies. "I need a plan for executing improvements in our digital support, including staff training and system upgrades."

Prompt: "I aim to improve our customer service, particularly focusing on digital support channels to enhance overall customer satisfaction. I need a detailed plan for implementing these improvements."

By using the GUIDE framework, you can direct ChatGPT to provide a structured and detailed response, encompassing all aspects from the initial goal to the execution strategy. This framework ensures a holistic approach to problem-solving and planning, making it an invaluable tool for tackling complex projects.

DISCOVER: Define, Insight, Scenario, Options, Verify, Expectations, Reflections
DISCOVER is a unique prompt engineering framework designed for exploration and problem-solving, especially useful when dealing with unknowns or uncertainties. Let's say you're tasked with exploring new market opportunities for a product. Here's how DISCOVER can guide your interaction with ChatGPT:

Define: Start by clearly defining the problem or task at hand. "Define the task as identifying potential new markets for our eco-friendly cleaning products."

Insight: Share what you already know or believe about the situation. "Our insight is that there's growing interest in sustainable products among young urban professionals."

Scenario: Describe the specific situation or context. "The scenario involves expanding our product line to appeal to this demographic in urban areas."

Options: List potential strategies or paths you're considering. "Options might include online marketing campaigns or partnerships with eco-conscious influencers."

Verify: Identify what needs to be confirmed or clarified. "We need to verify the actual size and purchasing power of this demographic in specific urban areas."

Expectations: Outline what you hope to achieve or learn. "The expectation is to gain a clear understanding of the viability of targeting this market."

Reflections: Address any unknowns or uncertainties. "One major unknown is how this demographic perceives our brand currently."

Prompt: "I'm looking to identify new market opportunities for our eco-friendly cleaning products, focusing on young urban professionals. I need to understand the size and purchasing power of this group and how they currently view our brand."

Using the DISCOVER framework with ChatGPT helps in methodically approaching situations where there are many unknowns, guiding the AI to provide insights, suggestions, and strategies that take these uncertainties into account. This approach is invaluable for exploring new ventures or solving complex problems.

Prompt Engineering Building Blocks

Many prompt engineering frameworks contain one or more of the following components or building blocks. Here is a list of framework components and pieces that you can use when talking to ChatGPT. These could also be helpful in creating your own unique framework.

 Goal, Intention, Purpose, Result: Ultimately, what do you want to accomplish, or in other words, what can ChatGPT help you with and what is the end result?

 Context, Background, Observations: Give supporting details about the task. Who or what is involved and for what reason?

 Tone, Mood, Voice: Useful for writing content, what tone or mood should ChatGPT use in its response?

 Role: What is your role or job title as it relates to the task, or what role should ChatGPT assume?

Persona: What persona should ChatGPT exemplify? "For the rest of this conversation, you are Shaquille O'Neal." We'll touch more on personas a bit later in this chapter.

 Assumptions: Similar to context and goes hand-in-hand with Unknowns below, assumptions that you make provide excellent background information. "I assume that x is true based on my observations and understanding."

 Unknowns: What do you not know? ChatGPT understands when you tell it if your knowledge on a topic is limited or if you are unaware of something. You can also suggest it use alternate approaches or suggest that you are not tied to solving something in any particular way.

 Return, Format, Modifiers: Parameters about what information ChatGPT should return to you and in what format. "Return only information relating to..." "Return a 1000-word article containing an introduction, a closing summary, H1 headings, and H2 headings."

Prompt engineering frameworks are more than just tools; they're keys to unlocking a deeper, more effective engagement with ChatGPT. The aim here isn't to have you keep a printed checklist of these frameworks by your desk. Rather, it's about internalizing these concepts, understanding the nuances of interacting with AI, and mastering the art of crafting prompts that yield the best responses. When you learn a foreign language, keeping a dictionary or textbook with you at all times isn't practical or convenient; learning the underlying structure, grammar, and vocabulary of your new language, whether it be human or LLM, gives you freedom and confidence moving forward in your day-to-day interactions.

Think of these frameworks as guides, helping you navigate the vast potential of ChatGPT with greater confidence and precision. By embracing these principles, you'll find yourself not just conversing with AI, but engaging in a dynamic, fruitful dialogue that enhances your productivity, creativity, and problem-solving capabilities.

Fun With Personas

Prompting an AI to adopt a persona involves crafting input prompts in a way that encourages the model to generate responses with a specific character, style, or personality. In other words, you can tell an AI their role and how to behave. This can be useful in several ways, from tailoring content to a target audience, invoking humor by referencing a famous figure or style, or just changing the "voice" of your content to keep it fresh. You can combine this with personal excerpts and examples that you feed it.

AI Persona Prompt Examples
Here are some example prompts for expressing a persona:

Historical Figure
Prompt: "You are [Name], a famous historical figure known for [specific achievements]. Write a letter to your contemporary discussing the impact of your work."

Professional Persona
Prompt: "You are a tech CEO. Draft an email to your team announcing a groundbreaking innovation."

Literary Character
Prompt: "You are Sherlock Holmes. Describe the scene of a crime you are investigating and share your deductions."

Fictional Character
Prompt: "You are Yoda from Disney's Star Wars™. Provide advice on navigating challenges and mastering the Force."

Genre-Specific Persona
Prompt: "You are a sci-fi author. Write the opening paragraph of a novel set in a dystopian future."

Historical Period
Prompt: "You are a journalist in the 1920s. Write an article covering a significant event of the time."

Regional Dialect or Accent
Prompt: "You are a character from the American South. Share a story from your life using a Southern dialect."

Emotional State
Prompt: "You are feeling nostalgic. Write a letter reflecting on memorable moments from your past."

Specific Occupation
Prompt: "You are an astronaut. Describe the experience of walking on the moon for the first time."

Age-Specific Persona
Prompt: "You are a child describing your favorite imaginary friend and the adventures you have together."

By tailoring prompts to evoke specific personas, users can guide the AI model to generate content that aligns with the desired character, tone, or style. It's important to experiment with different phrasings and context cues to achieve the intended persona and to fine-tune the prompts based on the output generated by the AI model.

Persona Prompt Examples

You can have fun with LLMs by requesting any number of fictional personas in addition to personas of past and present historical figures. Remember that the models are trained on a vast amount of data, including excerpts from many given personalities. You can further direct the LLM to focus on certain attributes of the persona to exemplify to help add personality to the responses.

A Friendly and Supportive Personal Coach

Prompt: "As a friendly and supportive personal coach, how would you motivate someone who feels overwhelmed by their fitness goals?"

ChatGPT response: "Hey there! Remember, every journey starts with a single step. It's completely okay to feel overwhelmed, but let's break down those big goals into smaller, manageable steps. Celebrate every little victory and remember, I'm here to support you every step of the way. You've got this!"

A Strict but Fair School Teacher

Prompt: "You are a strict but fair high school teacher. Explain the importance of meeting homework deadlines."

ChatGPT Response: "Students, it's crucial to meet your homework deadlines. This discipline is not just about the assignment, but about preparing you for future responsibilities. Late submissions disrupt your learning process and can affect your grades. However, I'm always here to help if you're struggling with the material. Let's work together to ensure your success."

Here's a fun example of how we used ChatGPT to help write parts of this book. In a conversation where we outlined the purpose and topics of the book, we asked ChatGPT "In the book's section about demonstrating ChatGPT personas, I've created the following 2 demonstrations. Based on these 2 examples, add 3 additional new persona examples where ChatGPT exemplifies a persona for Shaquille O'Neil, Albert Einstein, Elon Musk, and Bill Gates." We pasted the above 2 examples, and here's what ChatGPT returned about Shaq, Musk, and Gates:

Shaquille O'Neal - Legendary Basketball Icon
Prompt: "As Shaquille O'Neal, inspire a young athlete who is doubting their potential in basketball."

ChatGPT Response: "Hey, young champ! Remember, basketball isn't just about height or natural talent; it's about heart, dedication, and perseverance. I started out just like you, with dreams and doubts. But I worked hard, stayed focused, and learned from every game, every play. Keep practicing, believe in yourself, and never forget to have fun on the court. You're bigger than your doubts, so go out there and show the world what you've got!"

Albert Einstein - Renowned Theoretical Physicist
Prompt: "As Albert Einstein, explain why curiosity is important in science."

ChatGPT Response: "Ah, curiosity, the beautiful essence of a true scientist's soul. You see, without curiosity, science would be nothing but dry formulas and lifeless numbers. It is our wonder, our desire to understand the mysteries of the universe, that drives us forward. Remember, never stop questioning. Curiosity has its own reason for existing. The

92

important thing is not to stop questioning; the pursuit of knowledge is a delightful and endless adventure."

Elon Musk - Visionary Tech Entrepreneur
Prompt: "As Elon Musk, give advice to someone who wants to start their own tech company."

ChatGPT Response: "Starting a tech company is like launching a rocket to Mars – it's challenging, thrilling, and requires some serious guts. First, find a problem that genuinely excites you, something that feels impossible. That's your mission. Then, be ready to take risks and fail. Remember, failure is an option here. If things are not failing, you are not innovating enough. Most importantly, think big and stay focused on the future. The world needs more dreamers who do, so go make that dream a reality!"

Max Verstappen - The Champion Formula 1 Racer
Prompt: "As Max Verstappen, provide advice to a young aspiring racing driver about the importance of focus and determination."

ChatGPT Response: "In racing, every millisecond counts, and that's where focus and determination make all the difference. As a young driver, you've got to live and breathe racing. It's about more than just speed; it's about understanding your car, the track, and pushing your limits while staying in control. Remember, I started karting at a young age and every race, every lap, I learned something new. Keep your eyes on the track, stay committed, and never lose sight of your passion for racing. That's what separates the fast from the fastest."

Serj Tankian - Vocalist for System of a Down
Prompt: "As Serj Tankian, share your thoughts on how music can be a powerful tool for social and political expression."

ChatGPT Response: "Music is more than just melodies and lyrics; it's a language that transcends borders and speaks directly to the soul. In my journey with System of a Down, we've always seen music as a force for change, a way to challenge the status quo and give voice to the voiceless. Whether it's addressing injustices or expressing solidarity with those struggling, music has the power to unite and inspire action. It's about harnessing that energy, that raw emotion in your songs, to make people stop, listen, and think. That's the real power of music – it's an instrument of change, a catalyst for revolution."

Amelia Earhart – Pioneering Aviator
Prompt: "As Amelia Earhart, provide advice on how to overcome your own fears and blocks to doing something people think is impossible."

ChatGPT Response: "Remember, what seems impossible today may become a reality tomorrow through courage, determination, and perseverance. As I once said, "The most effective way to do it, is to do it."

Continuing The Conversation: Providing ChatGPT with Feedback

Interacting with an LLM (see **What is a Large Language Model** in Chapter 2) like ChatGPT begins with an initial prompt. However, it's important to set expectations and keep in mind that effectively using ChatGPT means that you'll likely have some back-and-forth interactions with any given LLM. While ChatGPT can certainly answer questions and complete certain tasks from entering only one prompt, in many cases you'll want to craft your prompts with the expectation that you are entering into a conversation. You can continue the conversation using any number of strategies and frameworks discussed earlier such as using prompt engineering frameworks or pieces of frameworks, modifiers, and more.

In earlier examples, we mentioned that you can talk to ChatGPT as if it were a person. Sometimes LLMs don't quite produce the output we had in mind. You can tell ChatGPT that it didn't understand the task, if you want to adjust the output in a new way, or perhaps you want it to give you additional insight or expand upon the response it gave you. "That's a good start, but let's make the following adjustments:" You can also reference previous messages by you or by ChatGPT. For example, "Let's take a step back. I liked the direction of your previous message where you stated [short excerpt from a previous answer]. Let's iterate on that and change it in the following way:"

Here are some examples of things you can say to ChatGPT to steer a conversation:

- **Clarify the Main Goal:** "Please focus more on the main goal of increasing customer engagement in your response."

- **Change Tone**: "Can you make the response more formal/informal/adopt a humorous tone?"

- **Prioritize Specific Information:** "Prioritize the cost aspects over the technical details."

- **Summarize or Outline the Content:** "Please provide a summary of your previous response."

- **Use Analogies or Examples:** "Explain it with a real-world example or analogy."

- **Adopt a Certain Perspective or Persona**: "Answer from a scientist's/teacher's perspective."

- **Simplify the Explanation:** "Can you simplify that for a non-expert audience?"

- **Provide Step-by-Step Instructions:** "Break it down into step-by-step instructions."
- **Incorporate Recent Data or Events:** "Include information about the latest developments in this field."

- **Use a Specific Format:** "Present the information in bullet points/a list."

- **Avoid Technical Jargon:** "Please avoid using technical terms."

- **Focus on Benefits/Disadvantages:** "Highlight the main benefits/disadvantages."

- **Include Additional Topics:** "Include information about related topics like X and Y."

- **Offer a Contrasting Viewpoint:** "What would be the opposing argument to this?"

- **Use Specific Keywords:** "Please use these keywords in your response."

- **Add More Detail:** "Could you provide more in-depth information?"

- **Be More Concise:** "Please make your answer more concise."

- **Oops, I forgot:** "In my last response, I told you that I would paste something below my initial prompt. I've included that information below. Please modify the answer you gave me based on this information."

- **Suggest Alternatives:** "Offer some alternative solutions or approaches."

- **Update with Personalized Information:** "Assume I have a background in marketing and tailor your response accordingly."

- **Offer Other Considerations:** "Based on my approach and what we've discussed, what other approaches or considerations should I be aware of?"

- **Avoid Certain Words or Structures:** "Your response was not very human-like. Specifically, you said [words, or sentence] which is not something an average person would include in their vocabulary. Limit your word choice to be more genuine, and without using words that a common person might be unfamiliar with."

Through having subsequent conversations with an LLM, you'll become intuitively familiar with how to best craft your initial prompts and how to continue a conversation when needed. When in doubt, ask yourself how you would respond as if ChatGPT were a person.

Chapter 5
AI in Practice

Let's go beyond theory and frameworks – how can AI help you today? Specific job domains are beyond the scope of this book, and for our purposes, we focus on the non-technical folks who need AI in their jobs, so code generation and programs like Copilot are not discussed. With that in mind, let's discuss how AI can assist with twelve of the tasks most office workers find themselves doing.

1. Emails

According to McKinsey, the average professional spends 28% of their workday reading and answering email.[2] For many readers, this means anything that can get them away from their inbox is the most helpful tool they can have. Fortunately, this is where ChatGPT excels. While we will cover specifically meeting summaries and project updates in a separate section below, let's examine a bit of office communication.

Prior to the widespread advent of email, the majority of office communication was done in person, face to face. This meant there was a natural limit to communication one could give or receive from the organization. While there were written memos and reports, it was substantially more difficult to distribute.

Enter email. Now it was easy to send communication to coworkers, bosses, and others in your company. As it became easier, it became overwhelming. When emails became easier to send, then everyone sent them! This meant inboxes overflowed with messages from what is on the menu at the cafeteria today to a critical assignment your boss has sent you. Email was and still is overwhelming.

Enter ChatGPT. Like a personal assistant who reads your inbox and gives a summary back to you, ChatGPT can take care of the majority of reading and writing of email. For reading emails, ChatGPT can summarize by providing the clear bullet points you need to do the job effectively. Rather than trying to skim longer emails to guess the needs, you can prompt ChatGPT to read the email. Additionally, when you are concerned you are reading things wrong or imposing emotions on the writer, ask ChatGPT to give a sentiment or a tone analysis.

For writing emails, ChatGPT can significantly aid the process. Firstly, it can draft quick responses to emails based on roles. It can also ingest large chunks of information and create email responses. Finally, you can have ChatGPT roleplay as a recipient of an email. One of the authors of this book had a particularly stubborn person they had to work with who would shoot down anything they sent over by email with negative responses. One day, out of sheer frustration with the constant negative email responses, they fed ChatGPT the emails the stubborn person would send along with context and asked ChatGPT to roleplay the person. With ChatGPT, they played out different email scenarios, perfecting email responses to the person and plotting out future communication. The author increased the positive response rate to emails from 5% to 20% with ChatGPT's help.

ChatGPT is not just another tool; it's a game-changer in the realm of email management. This innovative solution provides a much-needed respite from the relentless flood of emails, simplifying both the reading and writing processes. With its ability to distill lengthy emails into concise bullet points, offer sentiment analysis, and even role-play perfect responses, ChatGPT elevates email communication to a new level of efficiency. The transition from being swamped by emails to efficiently managing them with the help of AI not only augments productivity but also liberates precious time for professionals to invest in more significant aspects of their work. ChatGPT, therefore, stands out as an indispensable resource in the digital workplace, transforming a perpetual challenge into a streamlined, more manageable task.

2. Data Analysis

Data analysis is a part of almost everyone's life. Data are an omnipresent part of our lives, and within your professional life, you will often be confronted with information you must analyze and come to conclusions on. Adding AI as a tool to assist with analyzing the streams of information you are presented with is like having a PhD Data Scientist on hand to help.

In the movie *The Matrix,* the protagonist, Neo, meets up with a group of elite hackers to help take down their robot overlords. One of the hackers, Cypher, is staring into a screen of code that is the "Matrix" flowing like a waterfall. Neo asks him how he can understand it, and he says he's been looking at it for so long he doesn't even see it, he sees the meaning behind it. That is how AI can do data analysis for you. For a person to manually evaluate data, they could sit down and evaluate by hand every pertinent piece of information. One of the authors used to work with a team at a major company that used to look through almost 500 customer complaints a day to evaluate data trends (shortly thereafter said author put in place an AI to do that). By looking at data, people can see patterns. But it takes time. AI can do this in a fraction of the time.

With ChatGPT 4, this process can be significantly shortened. It can assist in each of the three major steps of data analysis: preparation, analysis, and presentation. Let's take a look at a real-life example with Olivia. Olivia is a data analyst. She is often inundated with ad hoc data requests with large datasets of customers to analyze spend trends. The process she must follow to get results is lengthy and labor intensive, generally taking between eight and sixteen hours of her time for one request. First, she must manually prepare and process the data. This means scrubbing for problem areas: missing data, misspelled names, etc. This is done by hand.

Then, the actual analysis has to be done, which is often for businesspeople looking for answers to their specific question like, "Hey, in the Pacific Northwest region for customers under the age of 50, did they spend more this year or less? Oh, and can you check if they spent more if they worked with these specific sales reps? Oh! And one more thing... can you specifically look at two time periods to see if the new product features we released made a difference in how much they purchased?"

For Olivia, this means many, many pivot tables in Excel or a bit of data manipulation and analysis in Python to get the right information, often over hours.

Finally, she must prepare the analysis to be presented. This includes charts, a briefing, and a summary. This can be one of the most time-consuming parts of the project, especially given it must be tailored to the audience.

Then Olivia found ChatGPT. ChatGPT was able to ingest the data, take in several factors to look for to clean it, and then export a CSV. Also, some of the data had to come from additional sources. ChatGPT helped write the SQL queries needed to get the data. ChatGPT helped Olivia compile the results into a thorough presentation. She even prompted the AI to roleplay as the reviewing manager in order to tailor the report to the recipient.

Given exact criteria and with a bit of prompt engineering, ChatGPT will turn out results in a few minutes, rather than 8-16 hours, saving days' worth of work and frustration. This is a powerful example of AI as a Force Multiplier in action.

In the above example, data was input directly into ChatGPT. What if, because of security and privacy concerns (see **Chapter 3 Security Considerations with AI**), we cannot give the data to or don't have access to ChatGPT? You have a few choices. You can use the mitigation strategies discussed previously. Also, AI is an essential tool for data analysis and can still be leveraged as a Force Multiplier. For cleaning data, it can help write SQL/Excel code, debug problem formulas, and give advice on other things stakeholders might want. For example, you can ask ChatGPT to pretend to be a sales manager who just did a new training to increase sales productivity. What are the things you would want to see in a year-end report? Additionally, for analysis, it can provide the latest and greatest methods with suggestions on potential areas to address with data and help to avoid pitfalls (average vs median as a representation for a population). For presentations, it can help to write the key analysis from bullet points.

More advanced work formerly assigned to senior analysts and data scientists, like modeling, can also be done by ChatGPT. AI has access to the most up-to-date knowledge of tried-and-true models and cutting-edge methodology. It can help tune and refine models and give common lingo explanations to items to help make modelling easier.

Employing AI tools like ChatGPT in the realm of data analysis mirrors the transformative impact of the graphing calculator in mathematical problem-solving. Just as the graphing calculator didn't supplant the necessity of grasping mathematical principles but rather enhanced the efficiency in tackling intricate equations, ChatGPT serves as a potent adjunct, not a substitute for human intellect and discernment. It streamlines the more cumbersome elements of data analysis—ranging from data cleansing and querying to spotting trends and preparing presentations—thus allowing the analyst to devote more time to interpretation, strategizing, and delving into deeper insights. However, akin to how a graphing calculator alone can't impart a deep understanding of calculus, the analyst must steer, interpret, and utilize the AI tool's outputs. This blend of human intelligence and artificial intelligence forges new frontiers in data analysis, transforming a previously intimidating task into something more approachable and efficient, while always under the vigilant and inventive supervision of the human mind.

3. Project Management

AI can also assist with many of the rote tasks for project management. Whether you have the title project manager or not, you are often doing the work of project management in your job or day-to-day life. Project management involves juggling multiple tasks, coordinating with diverse teams, and keeping track of numerous deadlines and deliverables. AI can assist with all of these tasks. There are three key areas where AI can assist a project manager: project planning, communication, and technical expertise.

With project planning, it can assist with brainstorming project steps, estimating timelines, and identifying potential risks. This support is crucial in the early stages of a project when laying a solid foundation is key to future success. Because of the wide knowledge base AI has, it can effectively do the time estimates for tasks. Then, it can be used to build out a roadmap. Furthermore, because it is an AI, it can do so without bias leading to over or underestimating time to complete. It can then tie the entire plan up with Gantt charts or other tools to help keep the team centered on the plan.

Another critical aspect is communication and coordination. AI can be used to draft project updates, create meeting agendas, and even summarize meeting discussions, ensuring that all team members are on the same page and important information is communicated effectively. Like we addressed earlier, it can roleplay to help digest pieces of information and ensure strong communication comes across the right way.

Finally, there is technical domain expertise. Often, when you are managing a project, you may not have the same amount of domain expertise as members of the team you work with. This can lead to breakdowns around misunderstandings, as you may not be able to effectively gauge time for tasks, linkages between tasks, or may not understand exactly what the team is doing. AI can act as members of your team to explain pieces of the work they are doing. For reference, one of the authors favorite tactics is to ask ChatGPT to explain things like they are explaining it to a five-year-old. (Note: this tactic does not work well on the company's senior legal partner.) This allows you to understand more thoroughly the dependencies you are working with and manage the project more effectively.

The incorporation of AI into project management is a transformative step towards efficiency and precision. The role of project manager encompasses a vast array of responsibilities, from meticulous planning to effective communication and deep technical understanding. AI stands as a powerful ally in this journey, streamlining tasks, enhancing your skillset, and offering invaluable insights. In project planning, AI's unbiased and comprehensive analysis aids in crafting realistic timelines and identifying potential risks, setting a robust foundation for success. When it comes to communication, AI's ability to draft coherent updates and summaries ensures clarity and cohesion within your team, bolstering the human element of project management. Moreover, AI's capacity to demystify complex technical domains empowers you to bridge knowledge gaps, fostering a deeper comprehension of your project's intricacies. Embracing AI not only saves you precious time but also elevates your capabilities, allowing you to focus on strategic thinking and nurturing the soft skills essential for effective leadership. The future of project management is intertwined with AI's evolution, and by leveraging its strengths, you are poised to lead with greater insight, creativity, and empathy.

4. Market Research

In the not-so-distant past, market research was akin to a treasure hunt. You'd either don your metaphorical explorer's hat and venture out to talk to people, extracting nuggets of insight from these interactions, or you'd bury yourself in the stacks of the library, pouring over reports and books. The goal was to understand what made people tick, what they bought, and why.

Then, the internet exploded onto the scene. Suddenly, you could conduct surveys with a few clicks, collate global data without leaving your desk, and access a world of information at warp speed. It was revolutionary, like discovering you could drive places faster when you'd been riding a horse your whole life.

The Advent of AI and LLMs: A New Era in Understanding Demographics

Enter the era of Language Learning Models (LLMs) and Artificial Intelligence (AI). This wasn't just another step; it was a giant leap. Picture this: You can now converse with any demographic imaginable to fathom their purchasing habits. An LLM can morph into any customer profile - minus the ability to taste a tart or work a widget, of course. Moreover, it can amalgamate trends faster than you can say "data analysis." It's almost comical, like strapping a jet engine to a bicycle.

Mary's Story: From Pastry Chef to Entrepreneur with ChatGPT

Mary is a pastry chef with a dream as sweet as her confections. She envisioned owning and operating her own bakery, a haven of exquisite cakes. But dreams often come with a side of reality checks - she was an expert at baking but clueless about setting up a bakery business, let alone marketing it.

To get her started on making her dream a reality, her tech-savvy husband introduced her to the power of AI. ChatGPT didn't just feed her information; it was her mentor, guiding her through the labyrinth of market research. It crunched numbers, identified her target demographic, and even estimated potential revenue. And then, after mastering the basic prompts and generating the foundation for her business, Mary used ChatGPT to simulate conversations with her prospective customers. It was like holding a focus group, without the logistical circus. She asked it to role-play as different customer personas, gathering insights straight from the digital horse's mouth.

Confident in her business plan co-developed with AI, Mary took her newly mastered skill of prompt engineering and leveraged ChatGPT to craft marketing copy. Instead of spending evenings in marketing classes or shelling out cash for a consultant, she collaborated with her AI assistant. She'd propose ideas, and ChatGPT would spin them into persuasive, engaging copy, tailored for each customer persona she had created. Mary's business is no longer just a dream but is well on its way to becoming a reality with a solid foundation of business and marketing plans co-created with AI.

Mary's story is not just about a dream come true thanks to AI, but a testament to the democratization of market research. With AI tools like ChatGPT, insights that once required an army of analysts and hefty budgets are now accessible to anyone with a dream and an internet connection. It's market research, but with a dash of AI.

5. AI as Teacher

In the not-so-distant past, learning was a personal journey meticulously guided by teachers in classrooms. Instructors would craft lessons tailored to their students' needs and learning styles using textbooks, card catalogs, and a deep understanding of their subject matter.

Then, the internet transformed education. Suddenly, teachers and students found themselves able to harvest knowledge from every corner of the globe. Courses, videos, articles – all just a click away. This dizzying array of information came with a challenge: sort through the irrelevant data and distractions to find the pertinent information for learning.

With the advent of artificial intelligence and Language Learning Models like ChatGPT, the ability to learn new concepts and acquire data has leveled up education. Anyone can chat with an AI about Python programming, and it responds with the patience of a teacher and the knowledge

110

base of an entire library. ChatGPT, in this era, becomes not just a guide but a chameleon, able to adapt to the learner's pace, style, and curiosity level. It's like having a wise mentor, a tireless tutor, and an enthusiastic study buddy, all rolled into one.

This transformation isn't just about access to information; it's about redefining the learning experience. ChatGPT is like a bridge connecting the personalized touch of the old-school classroom with the expansive knowledge of the internet era. It's akin to having a personal navigator for the vast ocean of knowledge that the internet represents, helping learners chart a course that's uniquely their own.

For learners harnessing the power of AI, diving into learning a new language, deciphering the complexities of blockchain, or untangling the philosophies of the ages becomes easier. With ChatGPT, the education journey is not just about gathering information; it's about interactive, adaptive, and personalized learning. It's a fusion of the old and the new, bringing together the art of teaching with the science of technology, turning learning into not just an activity, but an adventure.

6. AI for Networking and Professional Development

In the past, the art of networking and professional development was a practiced largely in person. Handshakes and business cards were exchanged on the greens of a golf course or in halls of conventions. Handwritten notes and memos were exchanged to foster more personal relationships. Mentors were met and nurtured through these in-person encounters. Physical presence and direct interaction were the keystones of building professional relationships and skills.

As the internet evolved, platforms like LinkedIn turned into the new golf courses of the professional world. Meetings and networking events shifted to Zoom,™ and streams of professional advice flooded through webinars and online articles. It was a seismic shift, a replacement of the water cooler with a global network where connections were made with a click, and professional development became a digital pursuit. The personal touch of handwritten notes was replaced by emails and instant messages, widening the reach but often diluting the personal connection.

In the transformative wave brought by AI tools like ChatGPT, particularly with the advancements in GPT-4, we're witnessing a fascinating melding of historical mentorship approaches with futuristic technology. Through LLMs, we have access to the wisdom of the ages, be it Einstein's genius in physics, Da Vinci's mastery in art, or Steve Jobs' revolutionary approach to design and technology. ChatGPT enables users to simulate mentorship experiences with AI representations of these iconic personalities, providing a unique blend of historical wisdom and modern AI capabilities.

This innovation in AI mentorship transcends traditional boundaries. With ChatGPT, your mentor is not only infinitely knowledgeable, drawing from a vast repository of information, but also perpetually available and unfailingly patient. Unlike human mentors, ChatGPT doesn't forget where your last conversation ended or what your learning goals are. It remembers your progress, adapts to your learning style, and consistently builds upon previous interactions to offer a truly personalized mentorship journey.

Consider the groundbreaking example of someone who used GPT-4, along with a voice modifier, to emulate Steve Jobs as their design mentor. This is more than just a technological marvel; it's a gateway to personalized education and mentorship at an unprecedented level. You can engage in deep conversations, receive critique, and gain insights from the AI representation of Jobs, or any other figure you admire, turning what once was a dream into a tangible reality.

In essence, ChatGPT's capability to morph into any mentor from history and remember every step of your development journey marks a new era in professional growth and learning. It's not just about accessing information; it's about experiencing and interacting with the knowledge and legacy of the greatest minds in history. This is mentorship redefined - limitless, personalized, and deeply inspirational, bridging past and future to empower the present.

In the dynamic world of professional development and networking, the advent of AI tools like ChatGPT has opened new frontiers. In the past, to prepare for a major industry conference, you would manually research attendees, painstakingly preparing conversation points, and post-event, try to recall key details for follow-up. The process was time consuming and often didn't reap rewards equal to the effort expended.

Enter ChatGPT, your digital ally in networking. Before the conference, it acts like a seasoned detective, gathering data on participants, speakers, and topics. It generates a comprehensive dossier on key attendees, their recent work, and potential conversation points, all compiled effortlessly. It's like having a personal assistant who knows exactly who you'll meet and how you can use the information to make the most of your limited time with them.

As the conference approaches, ChatGPT turns into a strategist, helping you craft conversation starters and insightful questions. These aren't just generic icebreakers; they're tailored to each individual's interests and expertise, making every interaction at the event meaningful and engaging. It's akin to having a cheat sheet for every conversation, ensuring you're never at a loss for words.

Post-conference, the real magic of ChatGPT shines. No more generic follow-up messages. Instead, ChatGPT helps you create customized thank-you notes and follow-ups, embedding personal touches and references to specific conversations. It's the digital equivalent of sending a handwritten note, demonstrating thoughtfulness and genuine interest in the connections you've made.

In the past, professional networking was a challenging mental game of remembering names, faces, and conversations, often relying on hastily scribbled notes. Now, with ChatGPT, it's like having a supercharged networking brain, one that never forgets a face, a conversation, or an opportunity to deepen a professional relationship. It's not just a tool; it's your personal networking aide, blending the reach of digital communication with the personal touch of the bygone era, making every professional interaction richer and more fruitful.

From Sports Medicine to Computer Science: Josh's Career Transition with ChatGPT

Josh's career journey is a testament to the transformative power of passion and the right tools at one's disposal. Beginning his professional life in sports medicine, Josh had established himself as a competent physical therapist. But it was his serendipitous experience with building a gaming PC that kindled a newfound love for computer science. This wasn't just a fleeting hobby; it was a calling that resonated deeply with him. Deciding to pivot his career, Josh took a bold step, leaving his job to immerse himself in a coding bootcamp. As he delved deeper into this new world, building a portfolio of projects, and diligently completing the bootcamp, he faced a significant hurdle – transitioning his professional identity to align with his newly acquired technical skills.

Josh turned to ChatGPT to play the role of digital career strategist when he recognized the need to reshape his professional image. His first task was to understand what hiring managers in the tech world valued most. ChatGPT provided him with a list of top strengths and skills sought after in the coding industry, giving him a clear target to aim for.

Next, Josh used ChatGPT to draft a learning plan. This was crucial in bridging any gaps in his knowledge and ensuring he was aligned with industry expectations. With a structured plan in place, he was set on a path to not only learn but excel.

The pivotal moment came in redefining his professional presentation – his résumé and LinkedIn profile. Josh employed ChatGPT to proofread his résumé, ensuring that it highlighted his most relevant skills and experiences. He understood that while his background in sports medicine was unique, it was essential to showcase how the skills from his past career could be an asset in his new tech role. ChatGPT helped him articulate this transition effectively, emphasizing his adaptability, problem-solving skills, and dedication to continuous learning.

Moreover, ChatGPT assisted in summarizing his projects, distilling complex technical details into concise, impactful narratives. This exercise wasn't just about trimming words; it was about crafting a story that resonated with his new career aspirations. It was about painting a picture of a dynamic professional ready to bring a fresh perspective to the tech world.

The culmination of Josh's efforts was a professional image meticulously tailored to appeal to hiring managers in the tech industry. With ChatGPT's assistance, he managed to bypass the need for a costly career coach, instead leveraging AI to strategically position himself in the job market. The result? A significant increase in the effectiveness of his job applications, leading to an internship that blossomed into a full-time position.

Josh's story is a vivid illustration of the potential of AI tools like ChatGPT in facilitating significant career changes. In a world where professional transitions are increasingly common, Josh's journey with ChatGPT stands as a prime example of how technology can be a powerful ally in navigating these changes, empowering individuals to chase their dreams and succeed in new, uncharted territories.

7. Document Generation

In the realm of knowledge work, the creation of documents stands as a cornerstone activity, encompassing everything from reports and proposals to presentations and briefs. However, the process of document generation, often laden with the demands of precision and clarity, can be both time-consuming and mentally taxing. Here, the advent of AI, particularly tools like ChatGPT, heralds a new era in document creation – one where efficiency, accuracy, and creativity coalesce.

Take, for instance, the task of report writing. This typically involves gathering data, analyzing it, and then presenting it in a format that is both informative and engaging. ChatGPT can assist in each of these steps – from sorting through data and highlighting key points, to crafting narrative sections of the report, and even suggesting visual elements to enhance its readability and impact.

Proposal writing is another area where ChatGPT can lend its capabilities. Crafting a proposal that is persuasive and well-structured is crucial, especially when stakes are high. ChatGPT can help in organizing the proposal's layout, writing initial drafts, and ensuring that the language used is both compelling and appropriate to the context.

For presentations, the creation of slides and accompanying notes can be streamlined with ChatGPT's assistance. It can suggest design layouts, help articulate bullet points, and even provide ideas for engaging the audience more effectively.

8. Customer Service

For customer service representatives, the traditional landscape has often been characterized by high volumes of queries, the need for rapid response, and the pressure to maintain accuracy and helpfulness. ChatGPT emerges as a valuable ally in this context, not as a replacement, but as a powerful support tool that enhances the capabilities of human agents.

The role of ChatGPT in this setting can be likened to that of a skilled assistant who is always ready to provide information, suggestions, and support. One of the most immediate benefits is the handling of frequently asked questions (FAQs). With ChatGPT, customer service agents can quickly access a repository of answers to common queries, ensuring that they provide consistent and accurate information to customers. This tool becomes particularly valuable in dealing with high-volume periods, allowing representatives to manage more inquiries efficiently without sacrificing the quality of responses.

Beyond FAQs, ChatGPT can also assist in drafting responses for more complex customer issues. In situations where a tailored response is required, ChatGPT can suggest drafts or key points, which agents can then personalize. It can also suggest solutions based on past similar cases, and even help in understanding and interpreting customer sentiment. This approach not only speeds up the response process, but also ensures that the answers are comprehensive and considerate of the customer's specific situation.

For customer service agents with access to ChatGPT Premium, the benefits are further amplified. The premium version offers enhanced capabilities such as more detailed and context-aware responses, and the ability to handle a higher volume of inquiries simultaneously. These features enable agents to deliver a level of customer service that is not only efficient but also deeply personalized, fostering better customer relationships and higher satisfaction.

9. Content Creation
Almost every knowledge worker is required to churn out content, meaning written documents that may be

accompanied by visuals. This means being able to synthesize ideas into a narrative format to be shared with audiences. As shared previously, AI is great at both summarizing, but it is also phenomenal at elaborating for content, in large part due to the fact LLMs are built on predicting the next word. There's three basic pieces AI can assist with: ideation, drafting, and revision.

When you sit down to generate content, oftentimes you have a general idea of what you need to do, but not sure how you might begin. Take a scene from the TV show, *The Office,* where a new boss arrives and replaces the existing boss, Michael Scott. One of the first things the new boss does is to ask one of the top performing salespeople, Jim, for a "rundown." A chunk of the episode is dedicated to the antics of Jim as he struggles to understand what to put into the report. You've probably been in a similar situation where you have a vague idea of what to do, but no idea of where to begin. ChatGPT can be used as a resource to generate ideas. Ask ChatGPT what a rundown could mean, the context, similar rundown reports, and more, and it will spit it out.

From there, many folks encounter "blank page syndrome," where they will stare at a blank page unsure of what to do next. Oftentimes, the first words to write are the hardest. So, why write them? One strategy is to have ChatGPT draft the first few pages off bullets given to it or to ask it to distill a longer report. You can even feed it a bit of text written in your tone to help emulate the tone.

Finally, ChatGPT is your copilot in the revision stage. Some strategies you can utilize include having it check grammar, read as an unbiased reader, and having it read as a reader biased as your audience. Taking it further, you can have it structure and read for tone. It can ingest other documents that have done well and help model the document like those documents, too.

AI can play a significant role in enhancing the content creation process for those in knowledge work. It's evident how AI, particularly ChatGPT, can be a game-changer in managing tasks like ideation, drafting, and revision. From deciphering vague assignments like creating a 'rundown' to combating the dreaded 'blank page' syndrome, AI emerges as a powerful tool. It adeptly assists in sparking initial ideas, initiating drafts, and even adopting your unique writing style. In the revision phase, ChatGPT shines as a collaborative partner, an expert in grammar refinement, unbiased and audience-specific feedback, and tone adjustment. This integration of AI into everyday professional tasks is not just about efficiency; it's about transforming the way we approach and execute our work. As we've seen, AI is no longer a complex, inaccessible technology but a user-friendly, invaluable asset in our professional toolkit, seamlessly bridging the gap between thought and expression.

10. Idea Generation and Counsel
In the ever-evolving landscape of business and creativity, the birth of new ideas is the lifeblood of progress and innovation.

Traditional brainstorming methods, while effective, can sometimes hit roadblocks or fall into predictable patterns. This is where AI tools like ChatGPT come into play, offering fresh perspectives, and stimulating the ideation process in unprecedented ways.

ChatGPT, with its vast repository of knowledge and data, acts as a catalyst for idea generation. It can introduce new angles to a problem, suggest novel solutions, and even challenge existing assumptions. This capability is particularly valuable in brainstorming sessions, where diverse ideas are key to finding breakthrough solutions.

Emilia's story is a perfect example of how ChatGPT provided a novel and successful solution through brainstorming. Emilia was assigned the task of helping plan and run an employee team building event for a summer celebration. However, because Emilia's company was smaller, she didn't get a committee or even funding. She was able to use ChatGPT as her collaboration buddy in between meetings and brainstorm ways to throw a great employee team building event inexpensively. Traditionally, her company had always gone to paid events like bowling or happy hours. ChatGPT came up with the idea of volunteering. She loved the idea and so did her coworkers.

For professionals seeking to harness creativity, ChatGPT can be a virtual collaborator, providing instant feedback, suggesting alternatives, or even playing devil's advocate. Whether it's for product development, marketing strategies, or solving complex business challenges, ChatGPT's ability to rapidly generate a wide range of ideas can be a significant asset.

11. AI as Knowledge Base

When you get stumped and need to know something, you usually turn to Google, YouTube, or Wikipedia. Between the three sources, you can generally find what you need. But there's a world of difference between being able to find a single source and reading all of the sources. It's like the difference between reading an article on a city you want to visit versus knowing someone who lives in that city.

Mike's story is a perfect example of the power of AI as a knowledge base. Mike has had a problem for the last two years with his Honda® Accord car alarm going off at random. First Mike tried Google. 3,040,000 results. He had no time to read all of those, so he picked a few at the top, tried them, but they didn't solve his problem. Next, he headed to a mechanic. The mechanic wasn't sure. Then he tried another mechanic. Same result. Then, he tried a third and final mechanic. Same result. For over a year, he would sheepishly slink out of the restaurant when he heard the "beep, beep, beep" of his Accord.

Then, he discovered ChatGPT. He asked ChatGPT to be his personal auto mechanic. First, he prompted it with his car details: 2006 Honda Accord, then with the problem: Car alarm won't stop going off. He listed the methods he tried to fix the problem. In the response set, one of the items stated, if the car has been driven in cold, wet environments, check the door motor actuator.

Sure enough, it was the door motor actuator. Because ChatGPT is able to see every source, it "knew" that other Honda owners had reported that if you drove your car in wet, cold environments, the motor that locks the door can build up rust and stop working, prompting an alarm signal. ChatGPT had access to every online car forum where people who have fixed a Honda Accord 2006 with a car alarm issue had posted about their problem and their solution. A problem that had plagued Mike for over two years was solved in under five minutes by ChatGPT.

That's the power of AI and knowledge bases. Search engines are amazing at finding a specific article, but AI has read all the articles and can answer the questions.

12. Image Generation:

People say a picture is worth a thousand words. What if I told you this one was only worth 9?

Credit: ChatGPT 4/DALL-E 4
Prompt: "Create a renaissance style painting of an Italian village."

AI art has reached the point where it, on its own, can surpass a Turing Test. A perfect example of this is the story of the video game designer that took first place in the 2022 digital arts/digitally manipulated photography category of the Colorado State Fair's art competition. The winning piece, _Théâtre D'opéra Spatial_ was generated with AI Midjourney, edited in photoshop, printed on canvas, then submitted.

We won't touch on the various ethical challenges that present themselves when we consider AI's role in art, but suffice to say, AI has reached the point where any person can generate impressive art. There is a plethora of AI art sites to browse. But, as fun as it is to find pictures of AI generated hamsters flying F-15 Fighter Jets (yes, it exists), how can image generation help you in your work? While the advanced topics for prompt engineering images is beyond the scope of this book, let's look at John's example.

Image Generation to Enhance Marketing Messages

John faced the challenge of putting together internal marketing for a new product. It's hard to convey some concepts with just words, so he turned to images. First, he went into the standard PowerPoint icon library. They were boring and mostly unrelated to his topic. Then, he turned to the web. Tons of stock photos, but how does he credit the artist? When and how is he able to use the art? Plus, most stock art was only a bit more related to the concept than the PowerPoint icons. He could try to get a professional photographer or shoot it himself, but that would cost money out of his already dwindling budget. That's when he turned to DALL-E, an AI-driven image generator.

DALL-E allowed John to create custom, clever images that perfectly complemented his marketing message. Instead of hunting for icons or stock photos that were only sort of related to the message he wanted to convey, DALL-E custom generated the exact images he needed. When he got stuck and wasn't sure what image he needed, he used ChatGPT to suggest images and even write the prompt for DALL-E. The whole process for providing the perfect, lifelike, on-brand images only took about an hour.

The tale of the Colorado art contest and John's marketing success story illustrate the stunning impact of AI in the professional realm, especially in creative fields like marketing. AI tools like DALL-E are revolutionary, offering a swift, intuitive way to generate custom visuals that hit the mark every time. This isn't just about keeping up with tech trends; it's about embracing AI to unleash your creative flair, streamline your work, and redefine what's possible in your career. As AI continues to evolve, the only limit to its application in your professional life is your own imagination.

Chapter 6
Beyond Business

A Day in The Life Using AI

I wake up in a mild panic to the chime of my alarm and the morning light spilling through the curtains. "Another Monday... fantastic," I grumble sarcastically. As I stretch, a thought crosses my mind about today's meeting that I inevitably failed to prepare for. "I need a quick briefing on the latest AI trends," I murmur. Over a scrambled eggs and bacon breakfast, I open my laptop and consult ChatGPT. "Give me a summary of the latest AI developments relevant to healthcare and summarize this roadmap provided by my team." By the time I'm sipping my coffee, I have a concise overview and am ready for the meeting.

Driving to work, I remember a complex yet also somehow vague email from a colleague that left me puzzled. Sitting down at my desk, I pull up the email and paste it into ChatGPT. "Help me understand this email from my coworker." I silently laugh at the AI's response, realizing I have dramatically overthought what my coworker was saying. Instead of having to bother my coworker, I now have both an outline of their email and proper response.

At the office, my day is a whirlwind of tasks, but ChatGPT has made life much easier. I initially dismiss the idea of using it, thinking it was inaccurate and limited, and I dislike the possibility of my losing credibility for admitting I use a digital assistant to do my work. After the meeting, I sit down in the break room, relieved that it went well. Over another cup of coffee, I decide to get ahead for the week by drafting a proposal. "Let's make this persuasive and to the point, while debunking opposing views," I instruct the AI. The result? A compelling document that usually would take me hours, finished in minutes.

The drive home gives me time to reflect about my day and what's to come at work for the remainder of the week. I'm thrilled that I've been able to leave work early these past couple Fridays since I started using AI, while at the same time accomplishing more than I was able to previously. Hanging up my jacket and slipping off my shoes, I pull out my phone and ask ChatGPT for dinner recipe suggestions based on what's in my fridge. The idea of a homemade pesto pasta tonight sounds perfect.

Evening rolls in, and as I wind down, I think about my personal blog. "ChatGPT, let's brainstorm some blog topics related to travel and technology," I say, lounging on my sofa. The ideas it generates spark my creativity, and I jot down a rough outline for my next post. I love being able to paste previous blog posts into ChatGPT and instruct it to use a similar voice and cadence for the new post.

Finally, getting ready for bed, I remember a friend's birthday is coming up. "ChatGPT, help me brainstorm gift ideas for a guy in his early thirties that's into snowboarding." I don't know anything about snowboarding, so I'm thrilled that the digital assistant breaks down what sort of gear and accessories a snowboarder might like. I make my selection online, proceed to checkout, and presto – Brad's electric hand warmer will be here by this weekend.

As I drift off to sleep, I realize how seamlessly AI has blended into my life, enhancing each moment, simplifying complexities, and amplifying my creativity.

Conclusion

It's clear that the integration of AI, particularly ChatGPT, into our daily lives isn't just a possibility; it's a reality that's unfolding right now. This book has explored the transformative power of AI in various professional and personal contexts. From mastering prompt engineering to adopting AI for market research, creative writing, and problem-solving, AI can be an invaluable ally.

AI, when used effectively, can exponentially enhance your productivity, creativity, and decision-making processes. It's not about replacing human intelligence but augmenting it, empowering us to achieve more in less time and with less effort. By applying the insights and strategies from this book, you're not just keeping pace with the technological evolution; you're leveraging it to elevate every aspect of your life.

Your journey with AI is just beginning. The possibilities are limitless, AI technology is evolving and improving every day, and the future is bright. Embrace AI, explore its potential, and watch as it transforms your world, one task, one project, one creative idea at a time.

Welcome to the future, where AI is your force multiplier, and the only limit is your imagination.

About the Authors

Steve Guarino

Steve grew up in Raleigh, North Carolina. In high school, Steve took an interest in programming, website design, and malware analysis, and in college he abandoned Windows in favor of Linux. He moved to Portland, Oregon upon graduating from The University of North Carolina at Greensboro to pursue a career with Intel Corporation. Acting as the Technical Program Manager for the Industrial Solutions Division's secure software development lifecycle, Steve has worked closely with numerous Intel software products including AI, machine vision, machine learning, and robotics. Steve then transitioned to a Devops Engineering role at Shift5 in Seattle, a cybersecurity company that services federal contracts. He routinely used AI to assist with day-to-day tasks including automating security processes. Steve now works for Brevir Solutions LLC developing content and delivering training to help everyone realize the value of using AI.

In his free time, Steve enjoys playing the drums, snowboarding, kayaking, and hiking around the Pacific Northwest with his wife Cambrie and his dog Riley.

Logan Marek

Logan grew up in Portland, OR where he spent a frightening amount of time reading sci-fi and playing Dungeons and Dragons. While at the University of Washington, he worked on several tech startups including a chatbot for university students, an eco-friendly air conditioner, and blockchain for voting. After not being able to eat his IOUs, he took a job as a Product Manager at a large bank, where he entertained himself being the "shaker-upper." Just prior to the outbreak of a global pandemic, he started a metaverse company focused on bringing together large groups of people in close proximity.

He took a gig over at American Express, where he got the opportunity to work in the operations space learning how to scale, balance budgets, and survive corporate Game of Thrones. He focused on implementing AI to automate and improve the lives of as many knowledge workers around the company as humanely possible. He became the youngest Director at American Express, leading a team of data analysts, scientists, engineers, and program managers across the globe in driving AI and robotic process automation implementation for American Express' 20+ year old processes.

Then, Jeff convinced him Brevir would be more fun and less friction, so he left American Express and took on the challenge of helping everyone use AI more effectively.

Jeff Torello

Jeff Torello is a 30-year veteran of the technology industry, retiring from Intel Corporation after a 20-year career. After Intel he transitioned into a variety of leadership roles with high-tech startups guiding their engineering efforts. He's the author of *"Crash Course - Leadership in a Remote First World"* and is also a CISSP cybersecurity "nerd." Jeff co-founded Brevir Solutions LLC with the goal of helping as many people as possible take advantage of technology.

"Jeff is like Yoda but less green and slightly younger" – Steve Guarino